EVERYTHING HAS A
BEGINNING -
EVEN THE
UNIVERSE

EVERYTHING HAS A
BEGINNING -
EVEN THE
UNIVERSE

NICHOLAS P. GINEX

*A lover of humanity and a seeker of truth. I am here
to explore, to search, to study, to educate and to unveil.
While aerospace, engineering and science are my
fields of academics, yet history, philosophy
and humanities are my fields of research.
I desire to learn and teach a few.
So together, we
can seek the light and search for the truth.*

Library of Congress Control Number: 2020914334
ISBN: Hardcover   978-1-6641-2207-9
                 Softcover   978-1-6641-2206-2
                 eBook   978-1-6641-2205-5

Print information available on the last page.

Rev. date: 08/21/2020

To order additional copies of this book, contact:
Xlibris
844-714-8691
www.Xlibris.com
Orders@Xlibris.com
817297

ACKNOWLEDGEMENTS

*This book is dedicated to those
who have
devoted time and effort to
enlighten us with the truth.*

*The articles and books I have written have
come into fruition only through the
influence of many wonderful people who
have entered my life. In some ways,
they all had an effect on the thoughts
I am fortunate to impart to you.*

*The findings provided in this book
are for
millions of people worldwide.
It is time for the U.S. military to disclose the
greatest technical discoveries
of our era –
anti-gravity and zero-point energy.*

CONTENTS

LIST OF ILLUSTRATIONS

LEGACY OF A FATHER

The following is an extract from the historical
novel, *Future of God Amen*, Section 1.6 titled, A
Father Seeks to Reveal Truths to All.

**It would be derelict of this father not to reveal the truths learned
by extensive** reading and research; and by the exchange of ideas with
the many people that have entered my life. Truth can be elusive and
may take many years to comprehend based upon real life experiences.
This author has been fortunate to have come upon truths by accident
and, in many cases, by simply con-necting the dots through the
application of common sense. It would be a foolish gamble to wait
for somebody else to present the findings acquired in my lifetime.
Our lives are made up of too many different events that shape our
thoughts. Be they on an educational, social, and personal level, these
events combined with our intellect and sensitivity will always present
a different color of the way each of us see, interpret life, and develop
our thoughts for others to hear or read.

This father feels a deep responsibility to educate and prepare his
children for the world they live in. They were the initial motivation
for him to write his first book titled, *Legacy of a Father*. It was self-
published and its subtitle was, *The Evolution of God from the Past
into the Future*. Written over a two-year period it ended up with 643
pages, too long for the average reader. This book provided factual
information how the ancient Egyptians conceived one-universal God
and became the basis for many of his articles and books. To read
the book reviews and overviews of several books, you may visit the
website:

http://www.futureofgod.com

Regarding my articles, they may be viewed on the Internet site,

As a father who desired to inform his children of the traps and deceptive ideas propagated around us. He felt obligated to share thoughts that may enable others to get closer to the truths that he has earnestly tried to surface. It is the author's nature to be grossly offended when he or others have been made a fool of by means of lies and deceptive ramblings. However much it hurts, he prefers to always know the truth. He will not knowingly stand by and let his children and the public made to be fools. They, and all those who have the courage to examine new avenues of thought, will benefit by getting to know their own God and be less likely to end their spiritual search in disillusion and separation from God.

God may be your own personal god that is not bound to a particular religious faith. God is not conceived in the same way by each person and is formed by one's own sensibilities, intellect, acquired beliefs and knowledge.

This spiritual God this author alludes to is simply love for mankind. However you may conceive your God, you may be assured that He would rather have you love the people around you than to focus your love on Him. This author's responsibility as a father is to educate and prepare his daughters to make their own way in life. They need not bow down, prostrate themselves in a submissive manner, and humble themselves as if their father was a God. They need only to respect their father and mother for the love and precious time they have invested to help direct their lives, develop their potential, and become strong individuals who can stand on their own two feet. God may be present, but He cannot do what a father and mother can do for them. That is, to have them carry on the legacy of raising wonderful children in this world.

As mere mortals, we may never be able to know the whole truth about God. As a little boy, and throughout his life, he has always wondered who was God and where did He come from? To inform and educate people that the concept of one-universal God was conceived in the marvelous civilization of ancient Egypt, he wrote a paper titled, ***Provide History of Religion and God***. The paper provides a brief overview of how scripture evolved and became the basis for the religious beliefs of the major religions, Judaism, Christianity and Islam. The paper has been noticed by the Education Research and Information Center (ERIC), which is sponsored by the Institute of Education Sciences (IES) of the U.S. Department of Education. It has been posted on the Internet by ERIC and may be accessed via the link,

http://files.eric.ed.gov/fulltext/EJ1073192.pdf

Many of us have some doubts about the existence of God. Others find themselves with a desire to believe in God but unable to accept many religious teachings and traditional mores. There are others who feel that the concept of God is simply another form of philosophy that tries to find answers to questions of morality and the possibility of an eternal life. Then there are those who do not want to deal with the concept of God at all. They may simply accept that religion and belief in God is man-made to teach people within a community to respect and love one another. The greatest man of God announced three times a command, Love one another.[1] This command is more encompassing than, "Thou shalt love thy neighbor as thyself."[2] or, "It's nice to be nice."[3]

This book reveals the author's thoughts about the beginning of the universe, the phenomena of Consciousness, the existence of Extraterrestrials, and the discoveries kept hidden by the United States military through their control of the national news media, TV news,

---

[1] **Jesus Christ,** Holy Bible, King James Version, Gospel of John 13:34, 15:12, 15:17.

[2] **Holy Bible,** King James Version, Gospel of Matthew 22: 39

[3] **Anonymous**

movies, and publication of scientific and astrophysics papers. Every conclusion and assertion has been grounded in facts and references that have come from reliable sources. This author will not waste time on gobbledygook to prove a point. It is love for the truth that gives him the stamina to share his research efforts with you. There is no other agenda in this book than to open your eyes, widen your perceptions, and bring you closer to understanding yourselves, the God in which you believe, and prepare you for attaining knowledge, love and integrity that will enable you to one day meet other intelligent beings in the universe.

EVERYTHING HAS A BEGINNING -EVEN THE UNIVERSE

BRIEF OVERVIEW

How did the universe begin? Quantum physics has found that minute particles have evolved the atom, which is considered the basic building block of all matter. Physicists and scientists are investigating what appears to be an unknown force that is currently being characterized as Consciousness. This force exists at the lowest levels and appears to have a will to create inorganic and organic matter with a purpose to reach its highest evolutionary perfection – the creation of thinking organisms to articulate and comprehend its own Consciousness. It appears this will is instilled with a reproductive desire in all organic matter that fulfills a noble purpose, which is to love one another and all living things in our universe.

This book presents (1) very new ideas about the beginning of the universe, (2) new perspectives about Consciousness as a viable force of creation, (3) a viable alternative to the Big Bang Theory, (4) verifiable and substantial evidence that Extraterrestrials exist, (5) that anti-gravity and zero-point energy technology has been developed under U.S. top-secret programs, (6) spacecraft has been built and operational in the United States, (7) due to top secret operations of the new space technologies, America has lost more than five decades that would have decreased the threat of climate change and improve the quality of life for people around the world. Lastly, (8) all learning and religious institutions need to emphasize truth and integrity so that human beings will be accepted by intelligent beings in the universe.

ABSTRACT

An abstract is provided because it existed originally
as a paper. This paper has findings and ideas that
should be investigated by scientific and perceptive
minds. All ideas are worthy of reflection.

This author evolved a hypothetical idea that everything has a beginning, even the universe. It surfaced based upon a perception by Nikola Tesla that throughout space there is energy. The science of quantum physics has led scientists and physicists to conclude that matter is derived from energy. An engineer, physicist, and inventor, Paramahamsa Tewari, revealed that the electron, a fundamental particle, is created through the transformation of energy in space. His theory presents a space vortex structure used to calculate how energy in space creates the electron and mathematically derive many of its properties, such as its charge and mass. Physicists have documented that one or more electrons revolves around every atom and forms the elements of the universe.

The discovery that the electron is inherently associated with the formation of every atom prompts scientists, physicists and philosophers to understand the phenomenon of consciousness. For it appears that the electron is an intelligent master in the creation of atoms. Beginning with Nikola Tesla, scientists and aerospace engineers have learned to capture energy in space, which has led to new technologies, zero-point energy and anti-gravity. These technologies and the phenomenon of consciousness will be explored in this paper. They are presented to motivate scientists and innovative engineers to someday eliminate poverty around the world, solve the threat of climate change, and give mankind the ability to travel into interstellar space. Several questions will be investigated. Will mankind be able to transform their aggressive behaviors that initiate bigotry, hate, and violence, and to learn to love one another? Will mankind advance morally and intellectually with the new technologies? Will mankind's transformation to love one another become a reality so that they are welcomed by other intelligent beings in the universe?

Keywords: The Electron, Consciousness, Extraterrestrials, Big Bang Theory, Zero-Point Energy, Military Industrial Complex

INTRODUCTION

You are entering a scientific and philosophical
journey about the beginning of the universe. It is the
author's hypothetical belief that *"Everything Has a
Beginning – Even the Universe."*

Throughout life this author has reflected on how life began and
wondered who was God and where did He come from? Upon
retirement, books by religious scholars and Egyptologists provided
the benefit of learning about the written works of one of the
greatest Egyptologists, James H. Breasted and the highly acclaimed
archaeologist, professor James B. Pritchard. Portions of their written
works are produced in *Future of God AMEN*.

Translated hieroglyphics conclusively revealed it was the Egyptians
that conceived the first one-universal God, *Creator and Maker of
all that is*.[4] This God was revered and worshipped over 2,000 years
before the birth of Jesus Christ. His name is announced today at the
end of a prayer, supplication, expression of thanks and praise, and
worshippers even sing "Amen" in reverent tones. *Future of God
AMEN* presents the development of the Judaic, Christian, and Islamic
religions from the beliefs of the Egyptians that had an advanced
civilization as early as 4200 BCE. Also presented is an overview
of how Egyptian hymns were emulated in the Old Testament of the
Judaic religion, initiated in 950 BCE and finalized in 400 BCE. This
book is provided as a free read and may be accessed via the footnote
below[5].

---

[4] Nicholas P. Ginex, *Future of God AMEN*, para. 6.1.5.2, Amen, the God of
Creation

[5] Nicholas P. Ginex, *Future of God AMEN*, Chapter 6.0, Amen, the Universal
God, Pages 163-184; Egyptian hymns emulated in Genesis, Pages 170-172, Pages
224-228; http://iranpoliticsclub.net/philosophy/amen/index.htm

An illustrative overview of the development of Judaic and Christian scriptures from Egyptian hymns is provided in a written paper published by the Clute Institute titled, *Provide History of Religion and God.*[6] It was placed on an Internet on-line library by the Education Research and Information Center (ERIC), which is sponsored by the Institute of Education Sciences (IES) of the U.S. Department of Education. Sharing the Egyptian hymns was necessary to reveal how mankind came to believe in Amen, the first one-universal God. An historical understanding of how mankind came to conceive God is interrelated with trying to understand the beginning of the universe.

History allows us to learn how mankind conceived the beliefs in a soul, a hereafter, a Son of God (the pharaoh), and one-universal God. However, dedicated Egyptologists not only reveal how mankind came to conceive God, but their findings serve as a springboard to reflect on the beginning of the universe. Did it have a beginning? If so, what are the fundamental building blocks of matter?

Could the electron be the master particle that creates matter? In the creation of matter is there a consciousness that pervades the universe? At what level does consciousness exist to perceive thoughts? What are the technologies that enables the use of energy in space? Can the technologies that utilize energy in space be used for the benefit of mankind? Will leaders in government be courageous enough to take control of the new technologies kept secret by the military industrial complex? Can the national media, educators, government, and religious leaders come to realize they must transform the morality and character of mankind to advance with new technologies? Will people throughout our planet embrace knowledge, compassion and truth, so that they may someday meet other intelligent beings in inter-stellar space?

---

[6] Nicholas P. Ginex, **The Chute Institute**, Contemporary Issues in Education Research, *Provide History of Religion and God*, 2nd Quarter 2013, Volume 6, Number 2; **ERIC** Internet link: https://files.eric.ed.gov/fulltext/EJ1073192.pdf

It was after writing *AMEN, The Beginning of the Creation of God* [7] that this author acquired an interest in the possibility of Extraterrestrial life and began to reflect upon the above questions; this led him into the realm of Quantum Physics and stimulated the philosophical view that *Everything Has a Beginning, Even the Universe.*

There is much to present in an attempt to answer these questions. The answers will be provided with scientific facts that have emerged over the past century and in some cases, only the author's best hypothetical thoughts will be presented with the hope that they will stimulate others to continue the search for definitive answers. The author has no other mission but to inform and educate the public; for which reason, this paper not only provides his thoughts about the beginning of the universe but the foreseeable implications that need to be addressed. He has written two sayings that he lives by and motivates his writings, they are:

Knowledge is a wonderful Gift

Knowledge and love leads to Wisdom

Knowledge without action means Nothing.

It is not enough to investigate how the universe began, but also, what are the obstacles and behaviors that mankind must overcome to be successful in entering a new era and join Extraterrestrials in interstellar space. That is why contents of this paper is provided in four chapters with significant conclusions.

---

[7] Nicholas P. Ginex, *AMEN, The Beginning of the Creation of God*, Overviews and book reviews, http://www.futureofgodamen.com

1.0 THE BEGINNING OF THE UNIVERSE

During retirement, the research and writing of several books led the author into a relatively new science, quantum physics. It is the author's belief that the universe began with the electron. The electron is an intrinsic part of every atom that forms all elements on our planet and matter throughout the universe.

Figure 1 illustrates the basic structure of every atom showing it contains protons and neutrons and has one or more electrons that revolves around its nucleus. The atom is considered the basic

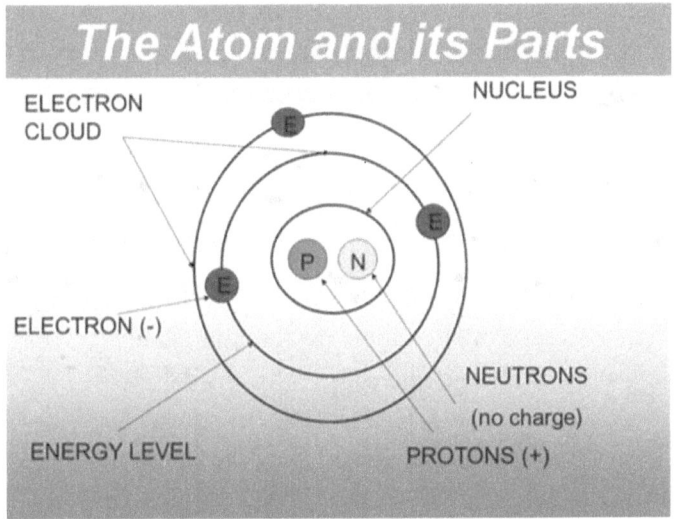

Figure 1. Every atom has one or more electrons.

building block of all matter throughout the universe.

A review of the Periodic Table of Atomic Numbers [8] verifies that the electron may be the mastermind in the creation of atoms. The

---

[8] **The Periodic Table of Atomic Numbers** can be accessed at: https://www.khanacademy.org/science/biology/chemistry--of-life/electron-shells-and-orbitals/a/the-periodic-table-electron-shells-and-orbitals-article

electron appears to have the inherent ability to coalesce protons and neutrons to form atoms that determine the characteristic of each element. Therefore, the electron may be the fundamental particle that creates matter. An Electron Configuration Chart[9] that shows the number of electrons for each element may be accessed in the Note below. There are 118 elements in the periodic table. Starting with the first element with one electron, each progressing element has one extra electron that revolve around the nucleus ending with 118 electrons for the last element, Ununoctium.

Of the many physicists who have discovered that the atom is made of yet smaller particles, an electrical engineer, who ventured into the world of quantum physics, developed a remarkable and profound theory that may unify the gap between macro and micro physics. This engineer, physicist, and inventor, Paramahamsa Tewari wrote a paper titled, ***Structural Relation Between the Vacuum Space and the Electron.***[10]

His paper reveals how the first fundamental particle in the universe, the electron, was created through the transformation of energy. He developed a space vortex structure that mathematically verifies how energy in space creates the electron and calculated many of its properties, such as charge and mass.

Tewari was dedicated to introducing to the scientific world a new scientific approach. He developed a space structure that allows deriving and substantiating the beginning of matter. This structure, named the vacuum-vortex, transforms energy in space and creates the electron.

---

[9] **Electron Configuration Chart**, view at: https://sciencestruck.com/electron-configuration-chart-for-all-elements-in-periodic-table

[10] Paramahamsa Tewari, ***Structural Relation Between the Vacuum Space and the Electron***, Paper accepted 10 February 2018; published 13 March 2018, https://www.tewari.org/uploads/3/9/2/2/39220475/16tewari.pdf

A photo of Paramahamsa Tewari is provided below as Figure 2. Figure 3 illustrates the space-vortex structure.

Figure 2. Paramahamsa Tewari reveald energy can be transformed into matter with his Space-Vortex Theory.

The space-vortex structure with a fixed volume of a dynamically stable void at its center is defined as the fundamental particle of matter, the electron. The properties of "electric charge" and "mass" of the electron and its "energy fields" are derived in Mr. Tewari's paper using mathematical equations associated with its structure. His paper presents a total new way that verifies the relationship between energy, mass and the speed of light. A brief description of Mr. Tewari's space vortex theory (SVT) is defined below.

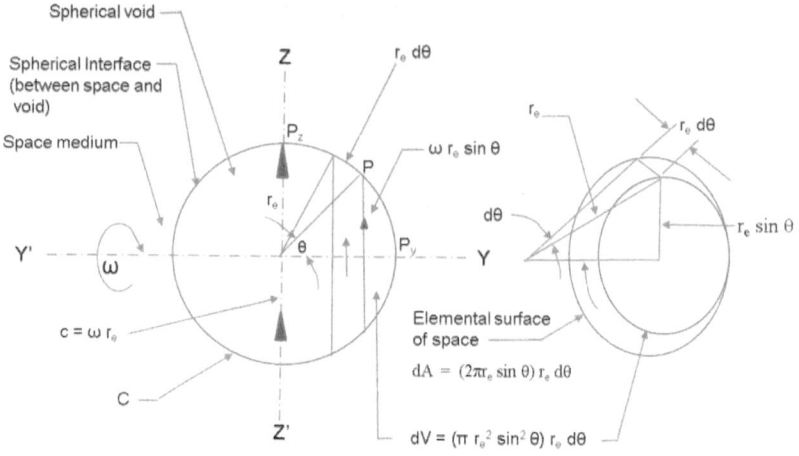

ω = Angular velocity of spherical interface around Y – Y'
Void = Fieldless spherical hole in space
Space is non-viscous, mobile, continuous, incompressible

Figure 3. Velocity field of the Electron Vacuum Space Vortex

vacuum-vortex. The conceived structure of the electron where its vacuum is emptiness and vortex refers to a whirling or circular motion that tends to form a cavity or vacuum in its center. Generation of electrostatic, electromagnetic, and gravitational fields are shown to arise from the vacuum-vortex structure of the electron. To verify his SVT, Mr. Tewari postulated three absolute principles:

1- The medium of vacuum space, throughout the universe, is an eternally existing, nonmaterial, incompressible, continuous, iso-tropic fluid substratum.

2- The medium of vacuum space has a limiting flow velocity equal to the speed of light relative to the fluid space (absolute vacuum).

3- The medium of universal vacuum space is eternal and inherent with motion. Vacuum space, vacuum, or space consists of non-material properties without matter and fields.

Tewari confirmed Nikola Tesla's memorable finding stated in 1891 that throughout space there is energy.

"Ere many generations pass, our machinery will be driven by a power obtainable at any point in the universe. This idea is not novel.... ....Throughout space there is energy. Is this energy static or kinetic? If static our hopes are in vain; if kinetic—and this we know it is, for certain—then it is a mere question of time when men will succeed in attaching their machinery to the very wheelwork of nature."

Tesla foresaw that "Throughout space there is energy" and Tewari concluded that, "the reality is that the dynamic space (energy substratum) is the first cause of creation, stability and subsequent existence of cosmic matter and fields." Tewari's book was published in 2009, titled *Universal Principles of Space and Matter, A Call for Conceptual Reorientation*.[11]

Figure 4. Nikola Tesla revealed there is energy in space.

---

[11] Paramahamsa Tewari, *Universal Principles of Space and Matter, A Call for Conceptual Reorientation*, Book Published 2007.

The physics world still needs to award the Laureate Nobel Prize to Paramahamsa Tewari for his scientific discovery of the creation of an electron and how the vacuum vortex structure is applicable to many fields of study in physics. Although he won early commendation by Nobel Laureates in physics and the Nobel Prize in Physics has been awarded 111 times, Mr. Tewari has yet to receive this distinguished honor. The scientific authorities would be remiss by not providing its greatest physics prize to Mr. Tewari in appreciation of his outstanding Space Vortex Theory.

This oversight palls to the ignoble fact that the Laureate Nobel Prize was never given to Nikola Tesla. Today, he is recognized as not only announcing that energy exists everywhere in space, but he is also the father of radio, television, power transmission, induction motor, the robot, and the discoverer of the cosmic ray. Tesla developed the alternating-current power system that provides electricity for homes and buildings. He also pioneered the field of radio communication and was granted more than 100 U.S. patents.

On November 27, 2017, Tewari died in his hometown close to Varanasi, Uttar Pradesh, India. He was looking forward to his 82nd birthday and completion of a new model of his reactionless generator, which he invented to validate his Space Vortex Theory (SVT). With the assistance of other engineers, Tewari verified the practical use of his SVT by building a generator that produced more power than it consumes with efficiencies of more than 200 percent.

1.1 The Electron Created the Universe

Mr. Tewari's paper reveals a new scientific breakthrough that electrons are born throughout space through the transformation of energy. This fundamental particle, the electron, had to be produced millions and trillions of times. In the same way, the vacuum space vortex may explain the formation of other fundamental particles, such as neutrons and protons. Through the supervision of the electron, many different configurations of atoms evolved to create the many elements of matter. Based upon the different configurations of an atom's electrons, protons and neutrons, the atom has earned the reputation of being the building block of inorganic and organic matter in our universe.

Professor David Tong [12], a theoretical physicist at the University of Cambridge who studies and lectures on quantum field theory, lectured at the Royal Institution in London and stated that the fundamental building blocks of matter are not discrete particles; instead, they are continuous fluid-like substances, spread throughout all of space. Physicists calls them quantum fields, which are electric and magnetic waves.

According to Professor Tong's recent theory of continuous fluid-like substances known as 'quantum fields' we find that even the electron and other fundamental discrete particles have a beginning. However, the professor appears to have made an over-reaching conclusion by stating that quantum fields are the real building blocks of the universe. Yes, quantum fields, energy in space, are from where discrete particles are born, but it is the particles that form into atoms, that are the building blocks of all matter. The question arises, is there an unknown force that the electron inherently possesses that drives the formation of different atom configurations, which creates inorganic and organic elements? It is this matter that creates the stars, planets, galaxies and dark matter throughout the universe not quantum fields. It should be noted that dark matter consists of trillions upon trillions of inorganic matter that are quite small and have not coalesced sufficiently to form perceivable matter.

What appears to be a phenomenon that requires further study is what is that unknown force that allows the electron to configure the many different elements in our universe? Is it consciousness? This phenomenon deserves further study and is presented below.

---

[12] Professor David Tong, *Quantum Fields: The Real Building Blocks of theUniverse,* https://www.youtube.com/watch?v=zNVQfWC_evg

1.2 Does Consciousness Pervade the Universe?

Having learned that the electron is one of the first particles to be created in space, it raised the question, does an unknown force, believed to be consciousness, cause the birth of the electron? For consciousness to exist, it must emanate from matter, a cohesive object, that is capable of originating thought. Since the electron is born from non-matter, energy that was able to transfer its electrical, magnetic and gravitational forces into matter, it may be hypothetically possible that energy created the electron to initiate the phenomenon of consciousness. Consciousness is non-matter like energy and it serves no purpose unless it is able to com-municate with matter. Therefore, could it be that energy created the electron (mass) so that consciousness can communicate in the universe? If so, with the creation of trillions of electrons through-out space, consciousness may pervade the universe.

The scientific discovery that positive, negative and neutral parti-cles of an atom can coalesce into inorganic and organic elements, gives credence to the idea that there is a phenomenon we may characterize as consciousness. Ultimately, we, as thinking human beings, may be the product of that source of consciousness. We are part of the "stuff" that makes up the universe, and in trying to understand our beginnings, it could be due to the inherent forces of the atom that surfaces as consciousness. This hypothetical idea that matter, created by atoms that determine inorganic and organic outcomes, can assume consciousness may have some merit because we are proof as thinking products of our universe.

When does consciousness actually become a reality whereby an element made of atoms is able to sense the external world and enable it to react to its environment and make choices? Logic presumes that reaction to its environment requires that the element cannot be an inorganic but an organic form of life. That is, we would not expect a rock to have the ability to react to external stimuli or think.

It would have to be an organic form of life that is capable to exhibit some form of consciousness. At what point in the development of organic life does consciousness reach a level that allows organic life to interact with its external world? For human beings, our first awareness (consciousness) starts after conception into conscious awareness at birth. However, a question surfaces as to why only at birth of an organic form would consciousness come into play?

Consciousness exists at many levels and in higher life forms provides the ability to communicate and exchange thoughts with others. But for consciousness to be productive and relevant, it relies on inputs of data and stimuli to construct thoughts and make decisions. Throughout the universe, Consciousness is like a baby that gradually becomes intelligent and articulate after evolving intelligent organisms, such as Human beings. After study and grooming from knowledge documented by dedicated writers, Consciousness reaches its height in human beings to cause them to think and reach out with hypothetical thoughts to try to understand its own existence. And that is why this author believes that the transformation of energy in particles is a long, evolutionary process whereby it transforms itself into matter with an inherent will to exist as consciousness in higher life forms that will question and examine its nature. Namely, thinking life forms are the crown achievement of consciousness in the universe.

Organic life, from a single cell to highly evolved animal and plant life forms, is preordained with conditions that support life. Organic life begins after inorganic matter (earth, gases and minerals) is exposed to heat and moisture that produces plant life, which produces a byproduct (air) to sustain life. Organic forms of life that become mobile, such as the fly, a bird, fish, and animals evolve based upon the environment and it all began with the mix of atoms that produced inorganic and organic matter. It becomes evident that the atom has forces that somehow takes advantage of its surroundings and gives birth to inorganic and organic matter. The forces within the atom have an intelligence or a consciousness that tries to evolve and express itself in many ways. It is this consciousness that reaches its height in human beings to think and eventually try to understand its own existence.

Anwar Shaikh, author, poet, philosopher, great Islamic historian and scholar, wrote a significant thought about consciousness; it complements the philosophical conclusion of this paper that, *"Everything has a beginning, even the universe."* Anwar rationally wrote,

"Since consciousness means cognition or knowing, there must be something worth knowing. Therefore, eyes have a multiple purpose; firstly, to play a definite role in the evolution of consciousness, and secondly, to know the world around us. From this conclusion, it also follows that the world or cosmos has a purpose: it wants to be known; it aspires to be conscious of itself. This seems to be the entire purpose of consciousness. Since man is the cosmic baby, he happens to be the medium for the universal consciousness."[13]

**Figure 5. Anwar Shaikh, the Indian Author,
Philosopher and Islamic Historian**

---

[13] Anwar Shaikh, *Mind and Matter, Cosmic Purpose*, Book 4, Chapter 12, http://www.iranpoliticsclub.net/library/english-library/eternity4/index.htm

Anwar Shaikh is to be applauded for his philosophical conclusions are compatible with the scientific findings that fundamental particles evolved the atom. It is evident from the latest quantum physics findings that energy waves are able to generate electron, proton and neutron particles that evolve atoms, which eventually, with an inherent consciousness, coalesce elements into matter that forms stars. We have learned though the astrophysicist, Carl Sagan, that stars can become unstable and explode due to gravitational forces that cause extremely high pressures and temperatures. The gravitational forces are so immense that astronomers have found it is responsible for pulsars, quasars, black holes, and star supernovas. These entities eventually reach a state whereby they will explode or emit matter, gases, and radiant energy back out into the universe.[14] This process repeats itself, which eventually populates the universe with stars, planets, and of course, dark matter.

1.3 The Big Bang Theory is Incompatible with Quantum Physics

The gravitational forces in matter verifies that it is *impossible* for a star to become a singularity that is so dense that it can create matter that could populate an entire universe. The pressure and extreme temperatures created by gravitational forces will cause the star, or singularity, to explode *before it* can coalesce billions and trillions of particles, gases and radiant energy that supposedly can initiate the birth of our universe. Therefore, from a logical perspective, the theory that the universe began with one big bang is not comprehensible. Rather, it is more likely that the universe has had multiple big bangs initiated by such phenomena as black holes, pulsars, quasars and supernovas. Truly, if particles, such as electrons, protons and neutrons populate the universe due to the transformation of energy waves, then there must be trillions upon trillions of particles that form atoms that

---

[14] Carl Sagan, *Cosmos*, **The Lives of the Stars**, Pages 238, 239

can coalesce into billions of stars and produce dark matter throughout the universe.

Another rational conclusion that the Big Bang could not have happened was provided by John Watson in an article titled, ***Top Ten Scientific Flaws in The Big Bang Theory***.[15] John reveals, "Since the Big Bang was supposed to occur only 13.7 billion years ago, then we should be looking at the early preformed universe. We shouldn't see fully formed stars and planets. However, instead we see stars and planets just like in our own galaxy. This is a serious problem for the Big Bang Theory because we're looking at the 'early universe' yet it doesn't appear very early at all. Thus, the Big Bang could not have happened."

An alternative to the Big Bang Theory, this author presents a philosophical view that scientists may eventually confirm; that ***"Everything has a beginning, even the universe."*** It is incomprehensible to believe that the universe began with the explosion of one singularity when in fact, it is constantly changing, dying, and forming new stars to exist. Just as inorganic and organic matter came to exist through an evolutionary process on our planet, matter has evolved throughout the universe through a consciousness inherent in particles. This author concedes that although the latest findings in Quantum Physics supports his philosophical view that everything has a beginning, scientists and physicists may never be able to explain how energy, in the form of electromagnetic waves, first originated. Einstein's equation, $E=mc^2$, presumes a balance exists between energy and matter. Still, for energy to create matter, in the form of particles, it had to appear first. The question remains for quantum physicists, how did energy in space first appear in the form of electromagnetic waves?

---

[15] John Watson, ***Top Ten Scientific Flaws in the Big Bang Theory***, (2015) https://thetechreader.com/top-ten/top-ten-scientific-flaws-in-the-big-bang-theory/

1.4 There is a Purpose for life to Exist in the Universe

Anwar Shaikh, a perceptive philosopher, has presented another aspect of consciousness. He wrote,

"If we delve deeper into consciousness, it transpires that it is the apex of evolution. Without it, existence or non-existence of the universe will not matter. A thing may exist but it is the knowledge of its existence which gives it a proper valuation. The universe obviously wants to be recognized, otherwise consciousness will have no meaning because whatever man sees, feels, senses or perceives relates to the universe; man himself is a part of it. Therefore, human consciousness belongs to the universe. More properly, the cosmos evolves man for the sole purpose of seeing, feeling, sensing and perceiving through him. Thus, man ranks as the cosmic baby with a special purpose. What is this purpose of man?"

The author's attempt to answer Anwar's question is with an imperative command given to us three times by a man of God – *love one another*.[16] It is through love that mankind will be able to reach the highest levels of integrity and truth that enables mankind to make sound decisions in the future. It is a command that has evolved with the highest regard for truth as illustrated by Figure 6. It illustrates Isis presenting the symbol of Truth to Nefertari. As presented in the book, *Future of God Amen*, it is the highest conceived attribute of the ancient Egyptians, their Pharaoh, esteemed councilors, and their revered God, Amen, the Lord of Truth.[17]

---

[16] Jesus Christ, *King James Bible*, **Gospel of John**, Verses 13:34, 15:12, 17

[17] Nicholas P. Ginex, *Future of God Amen*, **Amen the Universal God**, Page 181, 349. http://iranpoliticsclub.net/library/english-library/NicholasGinex-FutureOfGodAmen.pdf

Figure 6. Nefertari receives Truth from Isis. The symbol of Truth, the ankh, is the most highly esteemed Attribute of the Egyptian religion.

Truth depends on the ability to love one another, to listen with objective reasoning that sustains a compatible relationship with others. One may conclude that the attribute of love is inherent in consciousness for it instills the reproductive desire in all organic life forms to promote their existence. Can love for one another be the evolutionary purpose of consciousness? To attain the apex of development that is the jewel crown of being, the purpose of consciousness is to love all life throughout the universe. The undercurrent of consciousness is a profound will that instills the reproductive desire in all organic life forms, which promotes in our evolutionary development the attribute of love.

This apex or purpose of consciousness is challenged today by a religious ideology that has, since its inception, caused bigotry, hatred, violence and the killing of people. In the Qur'an,[18] the belief that Islam is the Religion of Truth and that Allah may make it prevail over every other religion has caused armies of Muslim fundamentalists to destroy many civilizations in efforts to subju-gate people to follow Islam. People around the world need to be conscious of this regressive ideology. They need to come together to educate misguided Muslims by revealing that the Qur'an needs to be revised by advocating the greatest command given to man – *love one another.*

Why and how people around the world can assist Muslims to develop the consciousness to love one another by revising the Qur'an is presented in, *Worldwide Communication Will Expose the Qur'an.*[19] It is provided as the last article of this book.

---

[18] Muhammad Zafrulla Khan, *The Qur'an,* **Sura Al-Tauba,** Chapter 9, Page 173, Verse 9:33

[19] Nicholas Ginex, *Worldwide Communication Will Expose the Qur'an.* http://www.nicholasginex.com/2018/02/07/stop-islamic-terrorism-by-exposeing-the-quran/

2.0. THE REALITY OF EXTRATERRESTRIALS

Before we explore why Extraterrestrial beings have entered our solar system and why they have an interest in human beings, we have got to reason if ETs (Extraterrestrials) and UFOs (Uniden-tified Flying Objects) really exist. After the downing of an ET spacecraft in July 1947 near Roswell, New Mexico, President Truman authorized a top-secret investigation by the military to assess the retrieval of dead beings that came from another world. This authorization quickly materialized into a highly secretive base identified as Area 51. It is a highly classified remote detachment of Edwards Air Force Base within the Nevada Test and Training Range and consists of approximately 60 square miles. Still active today, it contains hundreds of underground tunnels where scientists, biologists, and engineers work in isolated laboratories to study the anatomy of an Extraterrestrial and reverse-engineer the technology of an alien spacecraft.

Figure 7 is a photograph that was authenticated as being made in 1947 and therefore not something that's been photoshopped like today. This original photo shows an alien who's been partially dissected lying in a case. On the photo is a statement typed by the son of Alan Lewis who points out that in the bottom right-hand corner of the glass case is an Area 51 badge. It would be foolish to try to discredit the photo by claiming that the military man and alien was staged.

It is now 72 years since the alien crash and the American people are still fed misinformation by the military industrial complex (MIC). Although hundreds of UFO citing's occurred over Europe, England, Mexico, Washington DC and the United States, Canada, Australia, South America and countless other countries, the cover-up was effectively conducted by FBI, CIA,

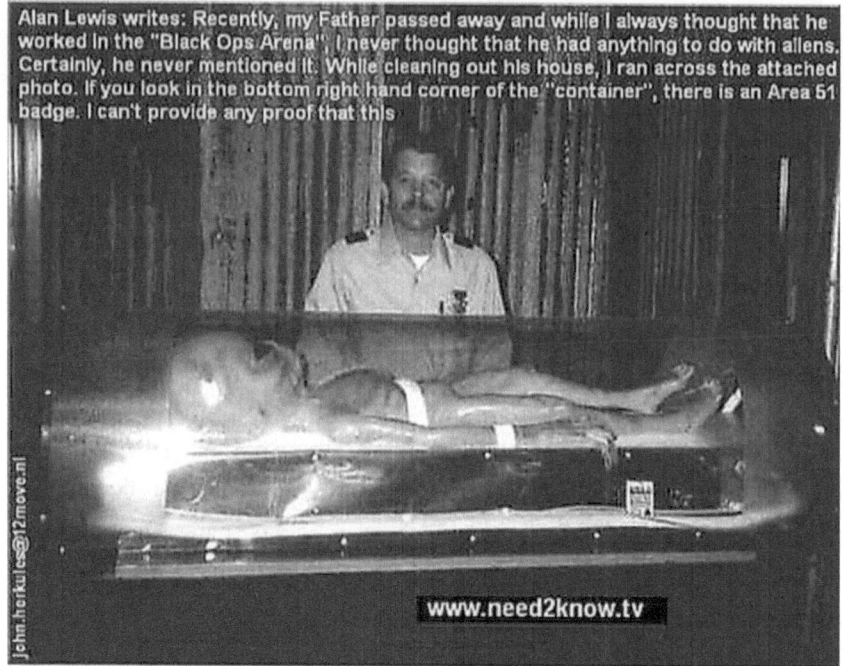

Alan Lewis writes: Recently, my Father passed away and while I always thought that he worked in the "Black Ops Arena", I never thought that he had anything to do with aliens. Certainly, he never mentioned it. While cleaning out his house, I ran across the attached photo. If you look in the bottom right hand corner of the "container", there is an Area 51 badge. I can't provide any proof that this

www.need2know.tv

john.herkules@12move.nl

Figure 7. An Extraterrestrial retrieved from the Roswell UFO crash.

DIA, NSA intelligence agencies and top DoD military personnel. National media control in the United States was achieved by controlling the national news media, scientific journals, TV and movie media financially; and by strict military orders to comply. This was an easy task as the MIC top secret projects, FBI, CIA, DIA, and NSA intelligence agencies were financially supported by U.S. Government tax payer dollars. The UFO-ET cover-up continues today.

Figure 8 presents the photograph of a female and male grey Extraterrestrial. The female is a rare photo that was provided by Robert O. Dean who spent twenty-seven years of active duty in the US Army where he retired as Command Sergeant Major after serving as a highly decorated infantry combat veteran. He also served in Intelligence Field Operations and was stationed at

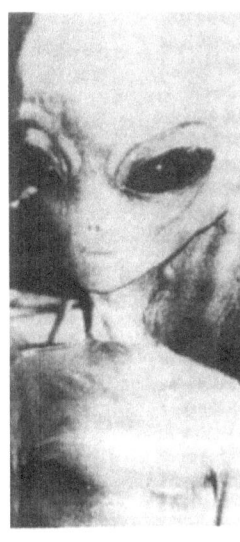

In the late summer of 1994, Robert O. Dean attended an international UFO conference in Mexico City. The photo on the left was given to him by a trusted and respected UFO investigator. Robert made known that he had always found the investigator to be honest and straight forward.

Being aware that disinformation and purposeful misinformation has flooded the field of UFO phenomenon, Robert cautiously posted the photo on the Internet. He was motivated to do so because he considered his source as well as authentic photographs he previously examined while in the military. Robert sincerely believes this photo to be a legitimate picture of an extraterrestrial (ET). This gray ET appears to be a female. On the right is a male.

Figure 8. A grey female and male Extraterrestrial.

Supreme Headquarters Allied Powers Europe (SHAPE), the military arm of NATO. As a retired Command Sergeant Major, Robert O. Dean's history includes forty years of research in the UFO field.

The national news media has successfully duped many Americans about the reality of Extraterrestrials and UFOs. Surprisingly enough, many Americans believe these phenomena are a hoax in spite of the hundreds of citings in many countries. It would be a poor reflection for those of us who have learned the truth from retired military officers and dedicated ET-UFO investigators to neglect revealing to all people the Area 51 top-secret coverup.

2.1 Zero-Point Energy Has Been Implemented for Space Travel

The MIC coverup has denied the American people the greatest discoveries of our era, anti-gravity and zero-point energy. Un-known by most Americans, these discoveries have been successfully implemented in U.S. spacecraft to enable interplanetary flight since the late 1960s. This author would be remiss by neglecting to provide the reader a presentation by the retired Command Sergeant Major, Robert O. Dean. Given at the Exopolitics Summit in Barcelona, 2009, he exposed the lies fed to the public about Extraterrestrials and provided an inspiring message of hope for mankind; view

https://www.youtube.com/watch?v=_
ngvIP0Za9M&feature=youtu.be

After the Roswell UFO crash, within one decade, a consortium of U.S. military and large aerospace corporations, namely the military industrial complex (MIC), developed and implemented with the knowledge of ET beings the use of zero-point energy and antigravity technology. The MIC has operated as a shadow government unaccountable to every U.S. President even though the U.S. government unknowingly authorized many billions of dollars in their Area 51 operations.

Today, unknown by the public due to national media control by the MIC, the U.S. military is able to dominate earth's outer space with Intergalactic Flying Object (IFO) technology. In 2014, Mark McCandlish, an accomplished aerospace illustrator, gave testimony at the **Secret Space Program Conference in San Francisco** where he revealed that the U.S. military not only has developed operational *antigravity*, but has for many years, developed and engineered *zero-point energy* that propels IFOs over the past 50 years.

Mark worked for many major defense contractors: General Dynamics, Lockheed, Northrop, McDonald-Douglas, Boeing, Rockwell International, Honeywell, and Allied Signet Corporation. Mark was previously in the US Air Force and for some of the major aerospace corporations was employed as an illustrator for aerospace designed spacecraft. He testified that his colleague, Brad Sorenson, with whom he studied and mentions at the conference, had been inside a facility at Norton Air Force Base, where he witnessed 3 different-sized alien reproduction vehicles, or ARVs, that were fully operational and hovering. They were all identical in shape and proportion, except that there were three different sizes. The smallest one was about 24 feet in diameter at the base. The next biggest one was 60 feet and the largest was about 130 feet in diameter.

Figure 9 illustrates the shape and much of the interior of the military ARV.

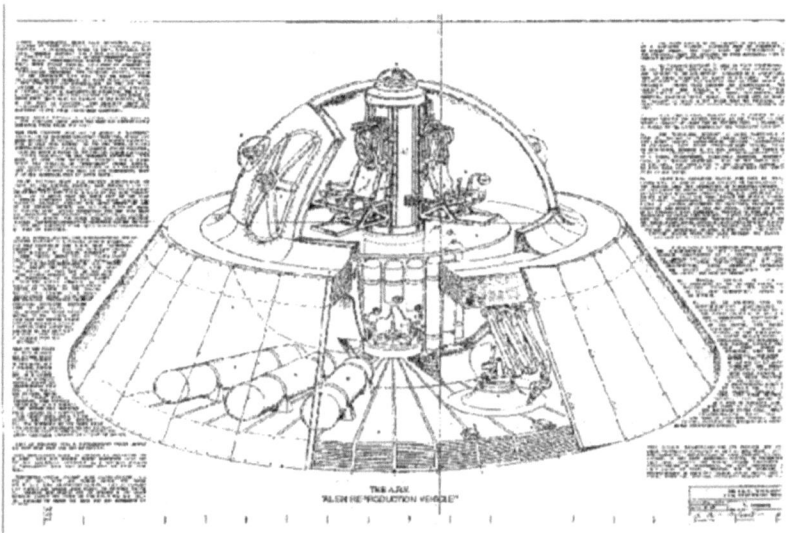

Figure 9. The ARV known to be operational before 1973.

Brad also saw a video tape that revealed the smallest of the three vehicles sitting out in the desert, presumably over a dry lakebed, someplace like Area 51. It showed this vehicle accelerated straight up and out of sight in just a couple of seconds with no sound, no moving parts, and no exhaust gases or fuel for propulsion.

A detailed description of the ARV and its components may be read in an article titled, <u>Mark McCandlish: Master of Aerospace Illustration.</u>[20] Mark presents the ARV on,

<u>https://www.youtube.com/watch?v=9QNvZN7X7v8</u>.

Note: The above video has been terminated.

A patent was filed by James King Jr. that looks like the ARV shown above except that instead of having a dome for a crew compartment, it has a cylinder in the center. The design has the same shape, the flat bottom, and the sloping sides. It has the coils around the circumference, and has the capacitor plates that are all radially-oriented. The patent was filed initially in 1960 awas secured in 1967, the same year that a photo was taken 52 years ago near Provo, Utah. It looks just like the craft shown as Figure 10. Kent Sellen, a former crew chief for the ARV, indicated the ARV, an interstellar flying object, was operational perhaps earlier then 1973, 46 years ago from this 2019 writing.

Having anti-gravity zero-point energy technology raises the ques-tion that government representatives should be asking, "Why spend all that money on development of NASA missiles for space exploration when the U.S. has technology that will take us into interstellar space right now?" It is clear that use of anti-gravity zero-point energy has not been revealed because the military industrial complex (MIC) has successfully kept these advanced technologies secret from the American public and government officials.

---

[20] Mark McCandlish, Mark McCandlish: Master of Aerospace Illustration, edited by Robert D. Morningstar. <u>https://ufospotlight.wordpress.com/2017/07/13/mark-mccandlish-master-of-aerospace-illustration/</u>

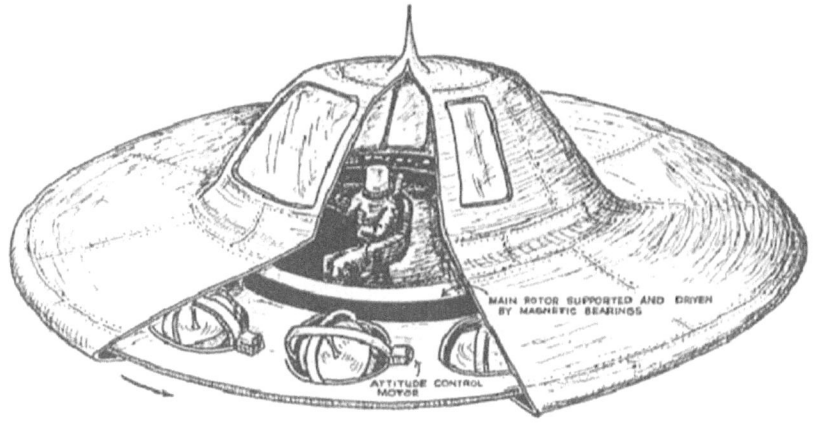

MARK II FLYING SAUCER

ELECTRONIC CENTRIFUGES BASED ON THE VORTEX DRIVE ARE MOUNTED IN GIMBALS TO TURN IN SYNCH
WITH THE REVOLUTIONS OF THE MAIN ROTOR DISC.
THE TUNED ELECTROMAGNETIC FIELD GENERATED BY THE VORTEX DRIVE CAUSES THE VEHICLE TO BE
CARRIED BY THE EARTH'S ELECTROMAGNETIC FIELD LIKE A DIRIGIBLE ELECTRON.
CONTROLLED GEOMAGNETIC PROPULSION IMPROVES THE DESIGN EFFICIENCY TO THE MARK III STAGE.

Figure 10. The Mark II Flying Saucer Prototype of the ARV.

It has been 72 years since President Truman authorized a top-secret investigation of a downed alien spacecraft by the military. It is time for the American people to loudly proclaim that the MIC disclose anti-gravity zero-point energy technology for commercial use to benefit mankind.

It is clear that the MIC has no intention to relinquish their power and control of the world economies as they continue to misinform and brainwash people to believe Extraterrestrials are a hoax and may be a threat to mankind.

2.2 Extraterrestrials Have Bases on Our Moon

The highly advanced technology of anti-gravity and zero-point energy is so profound that Extraterrestrials have been able to build a mother ship large enough to house hundreds of families and maintain dozens of small space craft. It is pertinent that we learn that Extraterrestrials are our neighbors. They have used the moon as a base for 8 decades and perhaps much longer. In Germany, two astronomers, Fred and Glenn Steckling, father and son, wrote about extraordinary findings in their book titled, *We Discovered Alien Bases on the Moon*.

Published in 2012 by the International Journal of Modern Physics and copyrighted by the World Scientific Publishing Company, this book was translated from German in 2005. Fred and Glenn spent many nights at the telescope looking at the moon and made an astounding find; a mind-blowing event, which has not received wide acclaim by the American media.

In Chapter 3, *Behind the Secrets of the Moon*, Fred Steckling wrote what he and his son saw on the moon crater Archimedes. During the month of November 1970, they used a 12.5-inch reflector (31.75 cm) to observe the field of Archimedes, which was 2,300 meters above sea level. One night they saw to their surprise three very large, cigar-shaped objects on the base of Archimedes. All three objects were the same size, two of which were parked in the northern area, the other was located in the southern area. They then compared the objects with a map section of the moon of this area created by the Air Force, which showed that the crater floor was relatively flat with no indications of these objects.

Fred and Glenn observed the three cigar shaped objects for several hours in the crater. The diameter of the Archimedes crater is about 80 kilometers, about 50 miles wide. According to their survey the cigar-shaped structures were at least **32 miles long and about 2.8 miles wide**. Figure 11 shows the 50-mile-wide crater occupied with three cigar-shaped objects.

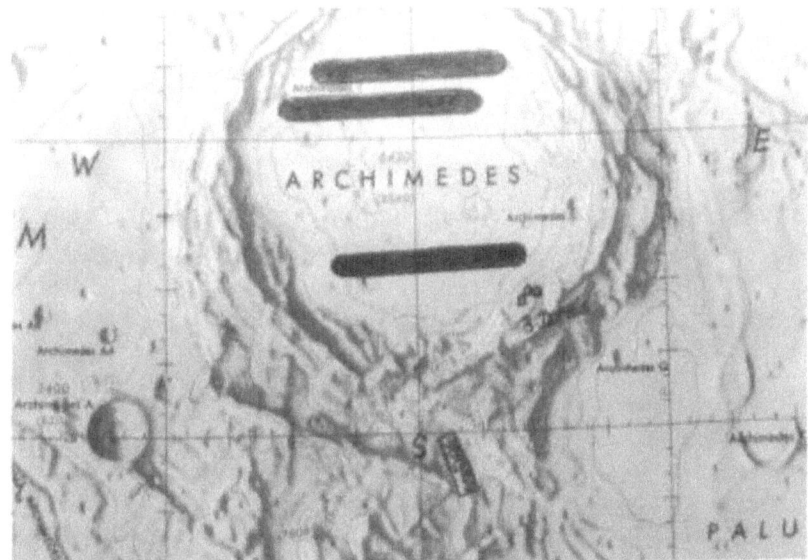

Bild 5 (Zeichnung): Gigantische Objekte im Krater Archimedes, wie sie vom Autor und seinem Sohn mit einem 12'/2-Inch-Meskop (31,75 cm) beobachtet wurden. https://archive.org/details/WeDiscoveredAlienBasesOnTheMoonGerman/

**Figure 11. On the crater field of Archimedes
were three Extraterrestrial motherships.**

The highly advanced technology of anti-gravity and zero-point energy is so profound that Extraterrestrials have been able to build a mother ship large enough to house hundreds of families and maintain dozens of small space craft. They have used the moon as a base for 8 decades and perhaps much longer. Steckling wrote,

"The very idea that someone has the knowledge to build spaceships of such monstrous proportions, is beyond the imagination. Photographic evidence, however, attest their existence. It would be useless to determine the material and labor costs incurred to construct only one aircraft several kilometers in length.

The construction of a huge aircraft carrier, about a quarter of a mile, cannot be compared to the immensity of the ET spaceship; a technical masterpiece, which would be extremely prohibitive

to build on Earth. A spaceship 32-miles long would be nearly impossible for humans to build."

Mr. Steckling astutely added,

"whoever they are up there, they are much more advanced than us."

3.0 DISCLOSE ZERO-POINT ENERGY FOR MANKIND

Out of arrogance and desire for power, covert leaders of the military industrial complex (MIC) have kept secret alien technology to surreptitiously control government, world organizations and wealth. It is a certainty that the MIC has retrieved aliens that have survived spacecraft crashes. It is plausible that they have communicated with aliens but the small-minded objectives of a surreptitious group and MIC leaders appear to only concentrate on world power and control. Will the United States government and people worldwide become knowledgeable enough to insist that all UFO ET spacecraft technology, developed covertly by the MIC, be disclosed to enter a new era that benefits every country by allowing commercial use of anti-gravity and zero-point energy?

It is imperative that people in all countries are able to employ zero-point energy to achieve low cost energy without the use of bio and nuclear fuels that are constantly polluting the earth. The benefits will allow people, around the world, to no longer endure poverty, lack of food, and gain leisure time to grow intellectually and morally. This effort by all people requires elimination of control by the MIC, a shadow government, which has denied accountability and control by the Congress, Senate, and Executive branches of the United States Government.

News media journalists and editors are educated and knowledgeable of the benefits of zero-point energy. As part of their profession, one would think that this highly advanced technology would be reported to the public to reveal its benefits for people around the world. But what is the problem, what stifles their ability to inform people worldwide about the benefits of zero-point energy? Could it be that their voices are stifled because the news media is financed and controlled by the military-industrial complex? The answer is that total control of the national media is financed with billions of American tax-dollars by subversive military control and multinational corporations. Zero-point energy is not disclosed because many industries do not want to change their production capabilities for fear that they will not be able to make the needed transition from coal, bio, and nuclear technology, which can result in a loss of jobs and profits.

After the Roswell UFO crash of 1947, President Truman authorized the military to conduct a top-secret operation to investigate the Extraterrestrial beings retrieved. Within one decade, a consortium of U.S. military and large aerospace corporations, referred to as the military-industrial complex (MIC), developed with the knowledge of ET beings the use of zero-point energy and antigravity technology.

Today, the U.S. military is able to dominate earth's outer space with Intergalactic Flying Object (IFO) technology. However, it is also clear that the military has operated as a shadow government unaccountable to every U.S. President by not sharing their IFO and ET discoveries. It is known that the IFO ET technologies developed by the (MIC) offers very powerful technological advances that can increase the quality of life for all people on earth. By using zero-point technology, poverty and food shortages can be eliminated and, by no longer needing bio and nuclear fuels the earth will no longer be subjected to pollution of our earth and all its life forms.

The shadow government that encompass the MIC and government agencies is so covertly pervasive that a serious warning by president Eisenhower was delivered to the American people in his 1961 speech. He stated,

"We have been compelled to create a permanent armaments industry of vast proportions. Added to this, three and a half million men and women are directly engaged in the defense establishment. We annually spend on military security more than the net income of all United States corporations. This conjunction of an immense

military establishment and a large arms industry is new in the American experience. The total influence -- economic, political, even spiritual -- is felt in every city, every State house, every office of the Federal government. We recognize the imperative need for this development. Yet we must not fail to comprehend its grave implications. Our toil, resources and livelihood are all involved; so is the very structure of our society.

In the councils of government, we must guard against the acquisition of unwarranted influence, whether sought or unsought, by the military industrial complex. The potential for the disastrous rise of misplaced power exists and will persist. We must never let the weight of this combination endanger our liberties or democratic processes. We should take nothing for granted. Only an alert and knowledgeable citizenry can compel the proper meshing of the huge industrial and military machinery of defense with our peaceful methods and goals, so that security and liberty may prosper together."

If President Dwight D. Eisenhower's speech is not unsettling enough for the American people, there is also a speech by the Honorable Paul T Hellyer, former Ex Minister of Canada, given in 2014 to introduce his book, *The Money Mafia: A World in Crisis*. His message affects millions of Americans who deeply care about the future of American and the planet earth. Identifying the U.S. is in grave danger, he addresses the iron veil of secrecy by a powerful group known as the Illuminati, the Cabal, or Industrial Complex. Paul Hellyer identified the *following entities that make up the New World Order.*

- *The Banking Cartel at the apex*
- The Oil Cartel
- The International Transnational Corporations
- The U.S. Intelligence Organizations FBI, CIA, DIA, NSA

Together, these entities control the American military industrial complex and the United States Media (news, book publishers, Internet, TV, movies and scientific, aerospace, physics journals).

The above bulleted entities and military industrial complex make up the **Surreptitious Shadow Government (SSG)** that has taken over the power of the United States. This mnemonic, SSG will be used in the rest of this paper. Readers are encouraged to watch the video by Paul Hellyer provided on the link,

https://www.youtube.com/watch?v=x-lDm0AitY0.

This dedicated man delivered a frightening message to all Americans on March 17, 2017. He emphasizes the insidious power the surreptitious shadow government has that leaves the president of the U.S. fearful of demanding exposure of the greatest technological discoveries of our era. His perceptive and ominous warnings may be viewed on:

https://www.youtube.com/watch?v=IWfKfuuj6Jo.

A top scientist, Rick Fousch, at the Naval Research Labs, the largest defense lab in the world, revealed that anti-gravity and zero-point energy was developed and implemented by October 1954. This means that by the mid 60's the capability existed to replace much of the oil, gas, coal, and fuel used for rockets, nuclear jets, cars, and internal combustion engines.

By the MIC not disclosing anti-gravity and zero-point energy for commercial use, people in America and many parts of the world have lost over six decades (2020–1954) of technological evolution. The MIC prevented the employment of zero-point energy because it would eliminate hundreds of trillions of dollars in oil and infrastructure assets that a surreptitious group controls and continually reap profits. To allow this group and MIC to maintain their grasp of global economies and the brain-washing of people through *fake news* (today an accepted and very popular phrase), the ominous warning by President Eisenhower has become a reality that threatens the well-being of mankind and viability of our earth.

There may be a solution with the authorization of a Space Force by President Donald Trump. It is obvious that the Space Force employs many MIC officers and personnel. It would be an astute president that has love for the American people to insure disclosure of anti-gravity and zero-point energy technologies for commercial use in all countries. Their implementation will usher in a new era of development that will benefit mankind both technically and morally. Although use of this technology would change the world by eliminating a centralized petrodollar economy, within a generation, it would energize the creation of many innovative industries that would increase the quality of life for all people.

3.1 The Military Industrial Complex Withholds Zero-point Energy

Dr. Steven M. Greer, an emergency room physician and Director of Emergency Medicine at a major hospital in North Carolina, left his medical profession to devote his life full-time to informing Americans of the truth about the reality of ETs and UFOs. About three years later he hosted a National Press Club Event in 2001 to initiate his disclosure campaign.[21] Many highly credible officers and scientific personal attended to give testimony that they witnessed or worked on UFO and ET top-secret programs. At the event, Dr. Greer proposed the following objectives:

- Open and honest hearings about UFOs and ETs in the U.S. Congress.

---

[21] Dr. Steven M. Greer, 2001 National Press Club Event, The Disclosure Project, May 9, 2001. https://www.youtube.com/watch?v=4DrcG7VGgQU

- That there be a permanent ban on the weaponization of space or the targeting of any objects of ET origin. To initiate such a ban, there must be immediate legislation and national/ international treaties to prohibit space based weapons.

- That there be a full and complete study of classified technologies connected to this subject to see how they could be properly declassified and applied for peaceful energy generation so that the world may get off of fossil fuels in enough time to prevent eco damage or war over the looming energy crises which is sure to sweep the earth in the coming decade.

- There must be declassification and release of currently classified technologies that could ameliorate the environmental and energy crisis. To eliminate poverty, improve the earth's air quality, and raise the quality of life for all people, we must make use of the anti-gravity and zero-point ambient energy propulsion systems that have already been developed on top-secret covert programs.

On January 23, 2009, Dr. Greer wrote on behalf of the *Center for the Study of Extraterrestrial Intelligence*, a letter for people to impress upon the President the need for disclosure. Attached to his letter was a Special Presidential Briefing (SPB) for President Obama and to his senior military and intelligence team. The SPB provided a full briefing, which contained detailed information on the projects, project code numbers, names, corporations, locations etc., associated with the UFO/ET covert operations. He wrote an appeal to friends, supporters of disclosure, to actively inform President Obama of needed actions regarding Extraterrestrials.

Nine months after the inauguration of a new and potentially transformative American President, we await significant progress on official Disclosure on the UFO/ET subject. While the UK, France, Denmark, Brazil and other countries around the world have increasingly opened their official government files, the US is found lagging behind her sister nations.

Again, on June 1, 2017, Dr. Steven Greer communicated with top administration officials of the newly elected President, Donald J. Trump. He hopes the new president will disclose and declassify ET UFO information once and for all. Over several years devoted to unravelling the secrecy of the military industrial complex, Dr. Greer claims to have evidence that a small clandestine group of people rooted in government has suppressed alien technology. Since the development and implementation of anti-gravity zero-point energy around the mid-1960's, the military industrial complex (MIC) has withheld this technology for commercial use; a period of more than half a century.

Dr. Greer's objective is to sensitized the public regarding Extra-terrestrials visiting earth. Many military witnesses, civilians and scientists involved with top-secret covert operations have given testimony of their encounters with ETs and UFOs. He has revealed that the Disclosure Project has coordinated interviews with more than 800 whistleblowers from high level defense contractors, government officials, military, and intelligence individuals to testify the existence of UFOs and ET beings.

Dr. Greer is dedicated to disclose the technologies that would usher humanity into a new economy and obtain the possibility to someday meet with other intelligent beings in the universe. He earnestly works to achieve such an endeavor and needs a tidal wave of support from the public.

However, even if President Trump is courageous enough to have the MIC disclose zero-point and anti-gravity technology that has successfully been developed for IFO interstellar travel for more than six decades, will the American people, and people around the world, be ready for the dramatic changes that will affect their lives? This question forces us to look to the future. It produces a bleak picture because an impasse exists between our political democratic and republican leaders, whereupon the Democrats desire of retaining power overrides the objective to create a better life for all our citizens. What needs to be pursued is that disclosure of such technological capabilities has the potential to benefit the economies of every country by development of new production capabilities, which will require an intellectual rebirth of educational and religious institutions.

3.2 Are Humans Ready for Intergalactic Travel?

There is an opportunity for humans to travel to intergalactic space and meet Extraterrestrial beings from other worlds. But are humans ready to meet ETs who are thousands and perhaps millions of years ahead intellectually and technologically? When we witness every day, the mass killing of people with different religious beliefs and witness the millions of people who suffer from poverty and lack of food it becomes clear that we, the human race, have not learned to solve problems that can eliminate religious differences, poverty and war.

How can ETs accept human beings that have not learned to live peacefully among themselves? We are a threat to ETs as we still let our arrogance, bigotry, greed for power and wealth, and hatred towards each other dominate. Our educational and religious institutions, around the world, have failed to unite their beliefs to teach integrity, an attribute that is based upon love for one another.

Even our national media have found it easy to lie about events and distort facts because of political differences. These differences have been infiltrated into the minds of teachers and professors whereby they propagate misconstrued facts to unsuspecting students. The alarming consequence is that in our elementary schools, colleges and universities young minds are poisoned with tainted political views by reinterpreting historic events. The President of the United States is ridiculed and disrespected with indecent names even though he has rescued our economy by increasing jobs and raised the incomes of blacks and other minorities. Unfortunately, it appears humans do not appear to be ready in two or three decades for intergalactic travel.

3.3 What Humans Can Do to be Accepted by Extraterrestrials

To initiate disclosure of IFO technology for public use, the President of the U.S. needs to provide Amnesty to all members of the military-industrial complex. A commission of scientists, engineers, corporate managers, business entrepreneurs, builders and city planners must study and design ways to employ IFO technology in all spheres of job production, such as manufacturing, agriculture, infrastructure and transportation. This planning may take a few years to design and develop technology that employs zero-point energy to implement new ways of building homes, growing agricultural products, increasing production of factories, and modifying land and air travel. The benefits will be the utilization of manpower and creation of jobs that will accomplish the building of new cities and modes of transportation.

The planning to turn our way of life into the era of zero-point energy will require incorporation of the new technologies in countries around the world. Astute and perceptive leaders from every country will understand that by employing IFO technologies their people will be relieved from poverty, jobs will open up, and opportunities for development of their people intellectually and morally can be achieved. However, there are fundamental problems that must be resolved. Educational and religious institutions that shape our lives intellectually, spiritually, and morally must be revised and restructured.

3.4 Zero-point Energy Requires a Rebirth of Our Educational and Religious Institutions

With the advent of an intergalactic era, followers of many religions may begin to question their fundamental beliefs. Many may become confused, ambivalent or angry to learn that other intelligent beings with very different beliefs exist throughout the universe. Such reality begs the question, are humans truly created in the image of God since other beings exist throughout the universe? Do Extraterrestrials have a spiritual belief in a creator God and if not, how do they function to establish harmony between their people and other intelligent life throughout the universe? Will the attributes of truth, justice, righteousness and integrity taught by inspired men who wrote scriptures and laws continue to guide the conduct of humanity?

Many people may become apprehensive of the beliefs they have been indoctrinated with but with honest reflection and knowledge of how their beliefs originated, they will begin to accept the fact that other life forms throughout the universe may also have belief systems of their own. The key to understanding any belief system is to acknowledge a basic attribute, which is inherent in the Consciousness of all living life forms – the desire to love and be loved. This fundamental aspect of living life forms was acknowledged by a man of God that gave us the greatest command – *love one another*. This attribute is part of everyone's human nature and is essential for happiness, peace and harmony between people of different races and beliefs.

Not having the good fortune of being taught how the various religions developed and not receiving an education to compare the beliefs of each religion, ignorance of people indoctrinated with different beliefs have and continue to cause bigotry, hatred, violence, and the killing of many innocent people with other religious or personal beliefs. Knowing that Extraterrestrial beings have existed for many thousands and even millions of years before development of the human race, it would be of great benefit to learn of the history, culture, and beliefs that have sustained their existence.

However, personnel who have worked within the Military Industrial Complex (MIC) for many years and actually interfaced with Extraterrestrials, have not inquired and documented anything that gives insights and understanding of their history, culture and beliefs. This has been a significant loss of an opportunity to learn about advanced beings and our universe. This is a sad that a surreptitious group that controls the MIC is more interested in developing weapons of destruction and domination of space around earth and intergalactic space and have no interest in elevating humanity with knowledge from Extraterrestrials. The limited objectives of the new U.S. Space Force place a veil of ignorance over the American people with fake news has kept them from learning how to live better lives and make a better world.

Ignorance still pervades the human race as a surreptitious group that controls the MIC and the fake news media, now is able to dictate the programs and activities of the United States Space Force. Yes, many good, honest men of integrity, unknowingly work for top-secret unacknowledged programs. But it is clear that the U.S. Space Force, created to protect America from global forces, such as China and Russia, has not specified the most important objective - to communicate and learn from intelligent beings that have achieved far greater cultural, technological and interpersonal development than humans. Why is this so? It is due to educational and religious institutions have not instilled the attribute of integrity and emphasized the command, *love one another*. As a result, integrity, honesty and truth have not been successfully ingrained in the upbringing of our youth and our military forces. Lies and fake news has become a common practice that distorts the ability of Americans to make sound decisions in a misinformed world.

It becomes a clear mandate for all educational institutions and religious organizations to take a look at their present teachings. History must be interpreted with clear facts that facilitates the learning process and, religious teachers have got to let go of outmoded beliefs. The Egyptians are a fine example of learning that scripture is not enclosed in concrete, never to be changed. After thousands of years of worshipping many gods, the Priesthood of Amon endorsed the vision of one personal god conceived by Amenhotep IV by evolving scripture that reveres one universal God, Amen. Figure 12 shows that beliefs do change as a people becomes more knowledgeable of themselves and the world they live in, albeit, the universe.

This author does not endorse any religion, they all serve their purpose to improve the morality and lives of its followers. However, it is to be strongly emphasized that all people should learn not only their beliefs but the religious beliefs of others. It is by comparison of religious beliefs that they become knowledgeable of the universal truth common to all beliefs. And yes, that common truth becomes clear as one gains knowledge – *love one another.*

For those of you who would like to take the journey that this author has travelled to learn how the Judaic, Christian, and Islamic religions have developed from the beliefs of the Egyptians, go to the website www.futureofgodamen.com. There, you will find overviews and book reviews that may interest you in learning what the author has researched and documented about the development of religion and belief in one-universal God. He harbors no agenda but to reveal what may give you other perspectives that will hopefully add to your spiritual beliefs, knowledge about our world, and prepare you for acceptance of other living beings in the universe. If interested in the overviews presented, free access to three books are provided on the Internet link:

http://iranpoliticsclub.net/authors/nicholas-ginex/index.htm

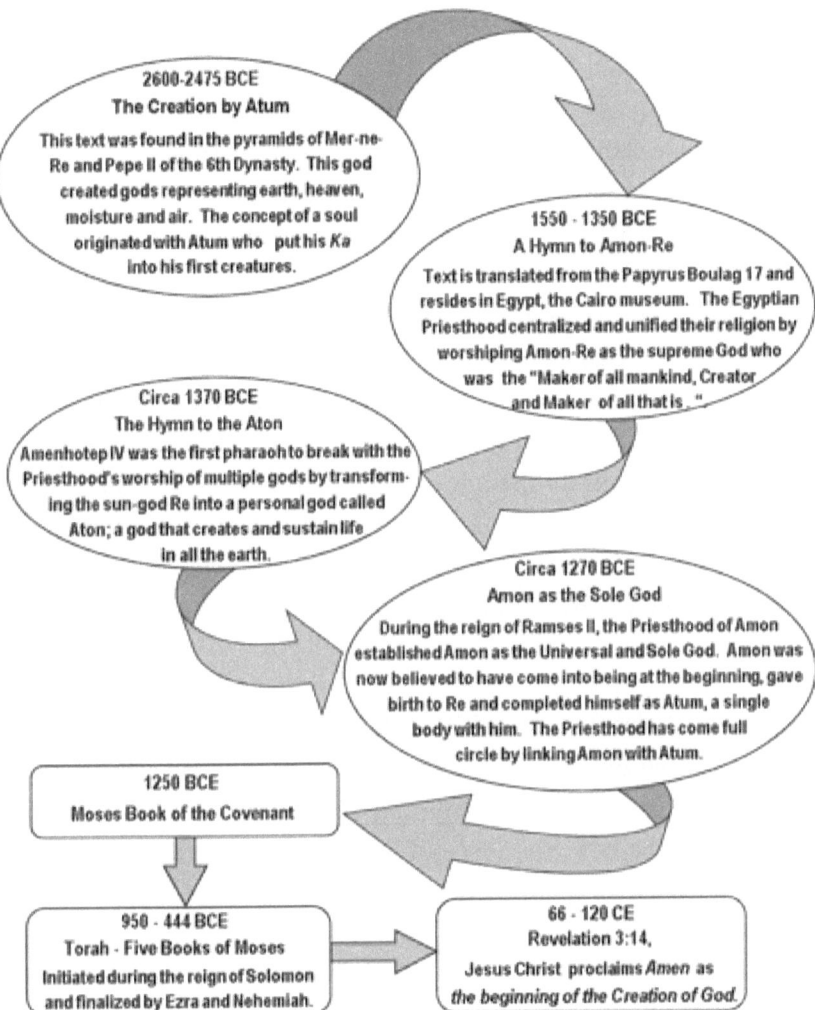

Figure 12. Scripture Evolves as Mankind Gains Knowledge.

3.5 The American People Must Insist on ET UFO Disclosure

With the creation of a U.S. Space Force, their first mission is to protect American space-based assets, such as satellites, against enemy attack and secondly, to deter a war in space by having the capacity to strike at those of an enemy that have developed weapons of destruction (Space News, Mark Whittington, June 19, 2018). Another mission is to clean up the debris left by dead satellites to ensure Earth space remains navigable. Looking forward to the future, the Space Force would provide rescue services, enforcing the law created for all space-faring countries, and arbitrate disputes among nations and private entities expanding their enterprise in the Milky Way and other galaxies. The Space Force can also provide the defensive capability against any threat from an asteroid that may threaten the destruction of our Earth and humanity.

Another article by space.com titled, *Space Force: What will the new military branch actually do?*, written by Leonard David, February 9, 2020 (https://www.space.com/united-states-space-force-next-steps.html), quoted the U.S. Secretary of Defense Mark Esper at a January news conference,

"It's just been recently that both China and Russia pushed us to the point where it now became a war-fighting domain,"

It becomes apparent that our U.S. military is positioning the Space Force for retaliatory and offensive capability against possible attacks from Russia and China. Mark Esper further emphasized the focus of the Space Force, which is to

"make sure that we're prepared to defend ourselves and preserve space."

Nowhere in the U.S. Space Force missions statement is there any reference to establish contact and communication with Extra-terrestrials. Is this a purposeful omission? If so, does it reveal that the Surreptitious Shadow Government (SSG) still intend to keep secret their retrieval of beings from UFOs and further development of anti-gravity and zero-point energy? If so, this portends a grave threat that the U.S. Space Force is controlled by the SSG and disclosure of zero-point energy will not become a reality soon. Instead of establishing an objective to learn about the history, culture and beliefs of Extraterrestrials, the focus of fear and distrust by our military leaders of the war capabilities of other nations can eventually lead to a catastrophic war that may destroy life on Earth. We have met the enemy, and they are us.

In the Space.com article by Mark Gubrud (identified above), a physicist and adjunct professor in Peace, War and Defense at the University of North Carolina raised some poignant questions,

"Will we continue this course toward destabilization and nuclear war, or will we renew our pursuit of arms control, disarmament, and the vision of a world free from this terrible danger?"

With an astute perception of what may loom in the future, Gubrud was motivated to state an appeal to initiate an objective for the U.S. Space Force,

"I call for renewed advocacy of space arms control: No war in space. No weapons in space. No weapons aimed at space. We need a global convention that says that."

The quandary still exists as to when will our President authorize disclosure of Intergalactic Flying Object (IFO) technology? The American people are educated enough to accept intelligent beings from other worlds. But will the arrogance and greed of the SSG continue to maintain the status quo of the establishment through fake news and control the destiny of mankind? It is essential that people be informed and educated to recognize that their political leaders must insist that the military industrial complex disclose IFO, anti-gravity and zero-point energy technology. Eventually, people will no longer tolerate being kept hostage from experiencing a new age that will increase the quality of life for all people and offer the experience of other worlds in the universe.

Zero-point energy can be employed in all moving objects and the needed infrastructure rebuilding of our cities, roads and bridges. Will people become informed and knowledgeable enough to want to increase their quality of life, meet Extraterrestrials and welcome an intergalactic era? Disclosure of IFO technology can be achieved by people writing their political leaders, Congressmen, Senators and the President to attain transparency to finally be free from insidious control that continues today by the surreptitious group and military industrial complex (See SSG in Section 3.0 after President Dwight D. Eisenhower's speech).

4.0 ZERO-POINT ENERGY FOR A NEW ERA

The human race is constantly evolving both technically and psychologically. Humans have advanced in the technical spheres, but due to the many diverse views about spiritual, religious, philosophical, and political beliefs, there is no common agreement in a belief system they can all embrace. As a result, much conflict has arisen that has resulted in hatred, violence and death of human beings in many countries. It would be of benefit to learn if ETs have solved their ideological differences and developed an acceptable view of learning how to love one another. This is the ultimate outcome; to revere all life and learn of the wonders of other beings in the universe. An alternate outcome is to let the fears and terror humans inflict upon other beings constrict our ability to advance to the next level of spiritual and intellectual development. If so, we may never get to fly into interstellar space.

If the military industrial complex discloses anti-gravity and zero-point energy will the American people, and people around the world, be ready for the dramatic changes that will affect their lives? The economies of every country will be impacted by new production capabilities, which will require an intellectual rebirth of our educational and religious institutions.

4.1 Will Humans be Accepted by Intergalactic Beings?

There is an opportunity for humans to travel to intergalactic space and meet Extraterrestrial beings from other worlds. But are humans ready to meet ETs who are thousands and perhaps millions of years ahead intellectually and technologically? When we witness every day, the mass killing of people with different religious beliefs and witness the millions of people who suffer from poverty and lack of food, it becomes clear that we, the human race, have not learned to solve problems that can eliminate religious differences, poverty and war.

How can ETs accept human beings that have not learned to live peacefully among themselves? We are a threat to ETs as we still exhibit arrogance, bigotry, greed for power and wealth, hatred and violence towards each other. Our educational and religious institutions, around the world, have failed to teach integrity, an attribute that is based upon compassion and truth. Even the National Media have found it easy to lie about events and distort facts because of political differences. Professors in our colleges and universities poison our young minds with tainted political views by reinterpreting historic events. The President of the United States is ridiculed and disrespected with indecent names even though he loves America and its people, has rescued our economy by increasing jobs, and raised the incomes of blacks and other minorities. Unfortunately, humans do not appear to be ready in two or three decades for intergalactic travel.

The root of the problem begins with behavioral instruction received by parents and morality instilled by religious leaders of spiritual organizations. They are most intimately involved in raising children with a foundation of moral rules that builds the attributes of character and integrity. This means that the scriptures of our religious institutions need to be evaluated to understand why and how they became a fundamental part of any civilization. Such evaluation begs to be studied because it is clear that it is the clash of different religious beliefs that are responsible for bigotry, hatred, terrorism, and the murder of millions of innocent people worldwide.

Religious instruction is essential to ingrain integrity, morality, and truth in the development of mankind. Introduced at the start of this paper, this author recommended reading, *Provide History of Religion and God*. The abstract of the Chute Institute paper indicates there is a need for high school, college, and university educators to introduce their students to a history of mankind's development of religions and beliefs in God. People need to understand that need and why instruction in morality is essential for the well-being of their nation and its citizens. By providing people with a greater understanding of the nature of man it can energize political and religious leaders to increase love between peoples of all nations and be able to welcome visitors from outer space.

4.2 The Covert Mind of US Operatives in Control of UFO/ET Projects

There are many theories about why Extraterrestrial beings are visiting our planet, and many have developed out of fear that these beings will invade earth. Such fear has been propagated by the MIC, CIA, FBI, DIA and NSA government agencies to obtain continued funding for the many underground complexes constructed to covertly study UFOs, recovered ETs, further anti-gravity and zero-point energy technologies, and perfect their intergalactic spacecraft capabilities that have been operational for more than five decades. The fears propagated by the military have even affected Air Force personnel to order its jets to shoot down any flying saucers that are detected on radar.

On July 29, 1952, UFO sightings were reported by the International News Service (INS) and revealed multiple flying saucers were seen along the East Coast and Washington D.C. This incident caused our military to give orders to shoot down UFOs. Such an order confirms the reality that covert intelligence agencies have been successful to utilize the media to not only alarm the public, but also our military personnel in believing that aliens were a threat to their lives. If there is any threat of war to humans by ETs, it will likely be initiated by our own military forces.

Public fear has been purposely generated by the covert intelligence agencies in order to continue developing experimental flying spacecraft. USAF IFOs (Intergalactic Flying Objects) were successfully operable and used to impersonate ETs in order to simulate alien abductions. This fear was also fermented by religious fanatics that work for and are in collusion with the covert government agencies. It is evident that many covert operatives and religious fanatics believe that public knowledge of another intelligence from outer space would mean a breakdown of our way of life. It is understandable that many people are indoctrinated with various beliefs in God and exposure to the beliefs of ETs may threaten their belief system. This imagined threat is fermented by military and religious leaders with the purpose to cause panic, turmoil, and cause people to reject Extraterrestrials from outer space.

There may be reason to believe that such highly religious people may be in control of the top-secret UFO ET programs. Possessed by such fears, men in control of the undercover UFO ET projects refuse to share their findings with our Government leaders and the American people. It is evident that military and intelligence agencies, in collusion with large corporations, have deliberately withheld facts about UFO sightings and examination of retrieved ETs, dead or alive, from the American people due to fear of the breakdown of existing religious beliefs.

The American people has been lied to by the MIC with use of worldwide news, TV and contrived movies, to instill misinform-mation and a perception that ETs are a threat to human beings. It is odd that after retrieving live ETs, so little is known or written about their culture, history, spiritual and worldly beliefs.

4.3 U.S. Military Downed UFOs and Killed ETs

Hundreds of UFO sightings were reported since the 1500s, and thousands more sighted after the 1900s. Bodily evidence, from July of 1947 through November 1992, is provided in a list of Extraterrestrials retrieved from UFO crashes in an article titled, *Understanding the*

Extraterrestrial mind.[22] Of 34 downed UFOs, a total of 137 ETs were retrieved and only 14 ET beings survived the crash. It has not been confirmed how many of the UFOs were immobilized by the military rather than ET loss of UFO control. The ETs did not threaten or fire upon any military sites, they were possibly surveying nuclear capabilities. The killing of intelligent beings from another part of our universe illustrates the arrogance and misguided leadership of our military forces that are not accountable to our system of government.

Dr. Steven Greer premiered his latest movie, *Close Encounters of the 5th Kind*, on March 4, 2020. In it he stated that,

"The military have targeted ETs around the earth and in space successfully killing many ETs and downing their spacecraft. Out here in Fort Huachuca, near Tombstone Arizona, there is an underground facility where there are 9 different ET craft that are there with their autopsy bodies."

This was reported through the men on his team that worked in the facility, near Tombstone Arizona. It is a depressing fact that our own United States military is responsible for killing intelligent beings that have come to visit Earth. Their technology must be advanced thousands of generations above ours to have travelled through space light-years away from suns within our own galaxy or a galaxy even further away, such as Andromeda that is 2.5 million light-years from our own Milky Way. With their advanced technology, they could have retaliated with deadly force to wipe out humanity and destroy our earth. But, thank God, they appear to have demonstrated that they not only have learned to live peacefully and love one another but have demonstrated the restraint and the intelligence to love the beings they encounter.

---

[22] Nicholas Ginex, *Understanding the Alien Extraterrestrial Mind.* http://www.nicholasginex.com/2018/10/30/understanding-the-extraterrestrial-mind/

It is apparent that our military leaders have demonstrated a level of arrogance, stupidity, and evil intent that surpasses the kind of violence we have been experiencing by Islamic extremists. The excuse of fundamental religious extremists is that they have been indoctrinated with an ideology that advocates the killing of innocent people that have a religious belief other than Islam. There are U.S. military leaders that are part of a greedy, powerful group (SSG) that want to continue control of not only the American people, but the world and beyond, which is interstellar space.

This paper warns all Americans and people around the world to "Wake Up" and demand disclosure of two of the greatest technological discoveries of our era, anti-gravity and zero-point energy.

4.4 Murder of UFO Investigators, Scientists and Engineers

The monstrous and despicable acts of killing Extraterrestrials pales with the number of UFO writers, scientists, engineers, and psychologists who have been murdered for wanting to reveal knowledge of Area 51 top-secret activities. A partial list has been extracted from an article titled, *U.S. Agencies Maintain Secrecy of UFO and Zero-Point Energy Technology* [23] and provided in the Appendix.

The MIC is responsible for more than 150 deaths of UFO researchers, biologists, engineers, scientists, and top military officers that had knowledge of, or worked in, Area 51. This list includes non-military people who had developed anti-gravity and zero-point energy capabilities or were authors, journalists, and radio/TV newscasters that wanted to reveal the existence of UFOs. Some deaths are suspicious while many others are outright murder planned

---

[23] Nicholas Ginex, *U.S. Agencies Maintain Secrecy of UFO and Zero-Point Energy Technology* http://www.nicholasginex.com/2020/04/26/u-s-agencies-maintain-secrecy-of-ufo-and-zero-point-energy-technology/

by a powerful surreptitious group within the U.S. military that used U.S. government agencies such as the CIA, FBI, DIA, NSA and DoD to conceal Area 51 activities.

The Appendix has been added to this book to apprise people around the world that all top-secret and unacknowledged projects in the United States military has got be under the control of our Government leaders. It is cowardice and a lack of leadership to allow the MIC to continue their insidious and surreptitious operations without any oversight and transparency of the greatest technological developments of our era, anti-gravity and zero-point energy.

Many ET remains and survivors are in secret underground compartments. Members of the White House, Congress and Senate have no control or accountability of the status of the ETs and technical UFO developments. They are blocked from all scientific and technical capabilities secretly developed in the hundreds of underground tunnels separated over many miles. In underground compartments, scientists and engineers who worked on ET UFO projects are isolated from one another so that no one scientific worker can access or acquire an overview of the entire program. The covert program objective was to develop spacecraft using anti-gravity and ambient zero-point energy that can enable travel in outer space greater than the speed of light.

The military, large corporations, and intelligence agencies worked together to convince Government leaders that contracts were necessary to continue UFO back engineering and scientific ET studies. Their pitch was that such funding was necessary to prevent nuclear war and an alien invasion. It has been estimated that covert, unacknowledged, top-secret projects cost tax payers as much as 40 to 80 billion dollars every year. The covert organizers of the UFO ET projects had the scientific objective to achieve anti-gravity zero-point energy to develop interplanetary space travel. Such energy, extracted from space, would make obsolete the use of oil, electric generators, and devices that run on bio fuels.

An unsubstantiated fear exists that the innovative use of abundant space energy would mean a total collapse of large corporations that employ thousands of people and affect the economic and financial balance of world economies. It became apparent to administrative and corporate leaders that such new technology may also cause panic if not taken in baby steps. But this is a short-sighted excuse for the new technology would unleash many new jobs and opportunities to build a new world that eliminates poverty, air pollution, raise the quality of life for all people, and permit them to have more personal time to advance morally and intellectually.

It is clear that the American people have been lied to in order to fund covert top-secret projects. When one considers that since 1951, as much as 20 to 80 billion dollars per year has been spent without their Government being apprised of covert UFO ET activities and findings, it is tantamount to fraudulent use of American taxpayer dollars. All of the covert projects could have been accomplished with transparency and truthfulness whereby our Government officials are given reports of the progress made by our UFO efforts and a greater understanding of ET viability.

Working in an open environment, the American people would have been exposed to the advancements made in the study of UFOs and ETs whereby, in gradual steps, humans would have become more knowledgeable and accepting of other living life forms. However, large corporations and limited minds, constrained with indoctrinated ideologies, took control of the covert top-secret programs and out of greed they seek to maintain the status quo and acquire greater power with the technologies of anti-gravity and zero-point energy capabilities.

Americans, and people in many other countries have the opportunity to obtain knowledge to utilize anti-gravity and zero-point energy whereby they can control their own destiny versus large corporations and corrupt-minded men dictating their future. They must loudly appeal to their leaders in Government to exercise their authority and truly make their country be run by the people and for the people; not the privileged few who only seek to gain power and satisfy their greed.

This paper is written to open the eyes, hearts, and minds of people everywhere to be informed and possess the knowledge that disclosure of the new technologies will benefit the world. The greatest lesson to be learned is that truth can set you free; for anything less, there will only be distorted minds, confusion, and eventually, destruction of mankind and possibly, our earth.

4.5 Reveal Extraterrestrial Culture, History, and Beliefs

It must be revealed to people around the world that ET life forms do exist and they must become aware of the benefits of communicating with them. Hopefully, by learning about the culture, history and beliefs of Extraterrestrials, mankind will learn of the higher purpose in life, which is to love all people and the living entities that are a part of our world and the universe.

4.6 Understanding the Extraterrestrial Mind

One questions why haven't ETs made personal contact with human beings? We must recognize that the intelligence of ETs is far above that of humans. Certainly, to have mastered technology to traverse millions of light years in spacecraft that utilizes anti-gravity and spatial energy, their development must be many, many generations, perhaps 100 to 1,000 times that of human life.

We must understand that ET beings could have destroyed many military facilities with weapons that surpass nuclear, atomic and hydrogen bomb capabilities. But they must have a high respect for the scientific and economic development of many of our civilized countries. They must also be aware of the many countries that are poorly managed with millions of people in poverty. Perhaps, ETs are watching our planet to see if mankind can rise above the many challenges on earth to improve the lives of all people. Surely, they are hoping that human beings can rise above the arrogance, bigotry, and violence precipitated by distorted ideologies.

As a visitor from outer space, you would be pleased to see another civilization has advanced to achieve a high level of technology and developed socially to produce literature that allows for continued growth mentally and spiritually. You would like to someday see that another life form has been able to advance and become another friend in the universe to exchange philosophies and ideas. It's like finding somebody else to play with; like a game of chess to enjoy the intelligence of another mind. It may be that it gets too lonely for ETs in the cosmos and to find there are other living beings is something to be thankful for. If you were an ET, would you not celebrate to have another life form with which you can exchange ideas and share your history with? Company with another life form is a wonderful thing.

Figure 13 illustrates the immensity of the Milky Way Galaxy and the position of our Solar System. It serves to help us realize that the universe has many billions of other galaxies, some even 10 times larger than our own. It will be a fascinating and wonderful experience to actually meet other beings in the universe.

**Figure 13. Position of our Solar System in the
vastness of the Milky Way Galaxy.**

If you were a visitor from outer space, you too would not interfere with the social, scientific, and spiritual development of other life forms. You would watch and see if the human species may someday learn to live with different ethnic peoples and unite with a spiritual ideology that was given to mankind more than 2,000 years ago.

Very simple, but very hard to achieve, a true man of love and peace gave mankind a command to be taken seriously, *love one another.* That one command is the key to survival of mankind; if successfully followed, mankind may also be able to join other life forms in the cosmos. You may inquire who that man of love and peace was by reading the ***Holy Bible***.[24] He was so emphatic, that he announced it three times not as a thou shalt but as a command.

---

[24] ***Holy Bible, King James Version***, (1972), **Gospel of John**, 13:34, 15:12, 17

It is clear that ETs do exist and is why we must try to understand the psychological makeup of the Extraterrestrial mind and what could be the motivation of ETs to learn about human beings?

It would be a marvelous breakthrough to learn about the history, culture, beliefs and knowledge of Extraterrestrials. To this day, we don't even know if they like to dance and sing.

5.0 CONCLUSIONS

1. *The universe began with the transformation of energy in space into fundamental particles, designated electrons, protons and neutrons.*

2. *Electrons revolve around all atoms and inherently possess a consciousness that forms the 118 known elements that are tabulated in the Periodic Table of Atomic Numbers.*

3. *Throughout space an infinite number of electrons create elements that populate the universe with planets, stars, galaxies and dark matter.*

4. *Like energy, consciousness is non-matter. Consciousness is a phenomenon that exists within the electron that facilitates the creation of inorganic and organic matter.*

5. *The Big Bang Theory is challenged with the findings of quantum physics that electromagnetic waves, energy in space, can be transformed into matter, which, over billions of years, populates the universe.*

6. *For 72 years, the military industrial complex has not disclosed to U.S. presidents the new technologies of anti-gravity and zero-point energy. They have misinformed the American people that Extraterrestrials are a hoax.*

7. *Since the late 1960's, the military industrial complex has developed and built space craft that uses anti-gravity and zero-point technology that can enter interstellar space.*

8. *Extraterrestrials have landed 3 mother spaceships about 32 miles long and 2.8 miles wide on the moon.*

9. *All people need to be informed that disclosure by the MIC of zero-point energy for commercial use will eliminate poverty, decrease use of bio and nuclear fuels, and resolve the worldwide threat of climate change.*

10. *To facilitate disclosure of zero-point technology, members of the surreptitious group, MIC and U.S. government agencies must be given amnesty for the murder of more than 150 persons who knew about Area 51 top-secret, unacknowledged programs.*

11. *Many corporations have created patents to suppress UFO ET technology for public use. Such patents are invalid for they were financed by billions of American tax dollars without government oversight.*

12. *The American people, and indeed people around the world, need to be educated about the culture, history, and beliefs of Extraterrestrials that have survived UFO crashes.*

APPENDIX. LIST OF PERSONS KILLED OR THREATENED BY THE U.S. MILITARY

No.	Partial List of Persons Knowledgeable of UFO Program Killed or Threatened by the U.S. Military 1 of 10
	https://www.youtube.com/watch?v=9QNvZN7X7v8
	http://ufoupdateslist.com/2003/jul/m10-009.shtml

1	**Robert Scott Lazar:** Physicist who back-engineered Alien Space-craft from another world for the United States Military at a secret base called S-4. In 1989 was a UFO ET whistleblower who stated he saw the operation of UFO and several aliens that came from another planet. CIA and FBI played mind games with him by moving objects around in his home and car. He said that the discovery of alien life could be the most important event in history.
2	**Eugene Mallove:** An American scientist, eminent physicist, founder of the New Energy Foundation, and expert in cold fusion who apparently had working tabletop devices was beaten to death in 2004 at the height of his career and the day before a public announcement of his findings.
3	**Stefan Merinov:** a Bulgarian physicist, researcher, writer and lecturer who defended his ideas of perpetual motion and free energy. It was reported that he committed suicide on July 15, 1997 by jumping from an outside staircase of a library building at the University of Graz.
4	**Mark McCandlish:** For exposing an illustration of the fundamental structure and components of the ARV, he had threats on the phone, the IRS came and took all his money, his cars, and he felt threatened for his life wondering why he hasn't been killed.
5	**Mark Tomlan:** A brilliant physicist and inventor of the patented Star Drive a system remarkedly similar to the ARV's in its use of zero-point energy and related space flight applications. He and Mark McCandlish corresponded numerous times before his unexpected death in his home in 2009, which occurred two weeks after the successful trial of a working prototype for a key component of his Star Drive system.

65

6   **Arrie deGeus:** An independent inventor, scientist, and theoretician who's not only rewritten the book on nuclear & particle physics, but has also used his theory's predictive powers to build a startling array of breakthrough energy technologies, including an IEC fusion generator and zero-point energy electrical generator. He invented a patented revolutionary clean energy technology associated with the zero-point field. He was found nearly dead in his car at an airport in 2007 and died a short time later. He missed his flight to Europe where he was about to obtain substantial funding for his work.

7   **Claudie Haignere:** In December 2008, the first French woman to go to space and a scientist who was supposedly at the forefront of alien-human DNA research, was rushed to the hospital after an apparent suicide attempt. She was rumored to have said, "Earth must be warned..." before slipping into a coma induced by an overdose of sleeping pills.

8   **Dr. T. Henry Moray:** Developed the Moray valve, a device for extracting radiant energy from the zero-point field and demon-strated this device hundreds of times and had hundreds of signed affidavits supporting his science. Yet in the end these two scientists were ignored and bullied. Dr. Moray's device was hammered down and broken into pieces by a competitor. Before he could piece it together he passed away from natural causes.

9   **Lester Hendershot:** Invented a magnetronic generator to energize an impossible flight if fueled by gas that took Charles Lindbergh in the Spirit of Saint Louis from New York to Paris. He fled to Mexico to continue work but was found dead at the age of 61 and it was attributed to suicide.

10   **Gaurav Tiwari:** Founder of the Indian Paranormal Society, died July 2016, under Paranormal circumstances. Only 32 Years old, recently wed and perfectly healthy. His death was ruled as 'a suicide'.

11   **CIA Director William Colby:** Died on April 27, 1996, of drowning in the Potomac River. His body was found 9 days later. The official report claims suicide.

12 **Stanley Meyer:** Developed and built an engine that runs only on water. After he applied for a patent, he got a call from two guys from the Pentagon. Stanley asked if he could release it for commercial use so that civilians could benefit from it. In 1996, Meyer was sued by his investors who claimed the device was not revolutionary despite his unique voltage intensifier circuit filed in the U.S. patent office. Meyer was brought to trial. Key evidence was not recorded and the microphone of the oral testimony was not made due to a malfunction. Later Meyer made an appeal and was found guilty of fraud and ordered to repay his investors. On March 20, 1998, Stanley joined his brother for dinner and two NATO officials for dinner. Stanley took a sip of cranberry juice, grabbed his neck, ran out of the restaurant and fell to the ground saying "they poisoned me."
Though he had documentation of his device, the full secrets died with him.
Dr. Greer and the Orion project in 2008, attempted to buy Stanley Meyer's science estate in a silent auction. However, they were outbid by an anonymous source.

13 **Gaurav Tiwari:** Founder of the Indian Paranormal Society, died July 2016, under Paranormal circumstances. Only 32 Years old, recently wed and perfectly healthy. His death was ruled as 'a suicide'.

14 **Max Spiers:** October 2016, he was found dead on a sofa in Poland, where he had gone to give a talk about conspiracy theories and UFOs. He had texted his mother, 'If anything happens to me, investigate' just days before his mysterious death.

15 **Philip K. Dick:** Science fiction author of Bladerunner and Minority Report for several years died of a stroke under somewhat mysterious circumstances on March 2 1982. He was writing a non-fiction book about his experiences with alien contact. It was never published, and the manuscript has disappeared.

16 **David Wynn-Williams:** World's leading microbiologist dies of a sudden and suspicious death. On March 24, 2002, died in a road accident near his home in Cambridge, England.

17   **Ron Bonds:** He published books on unsolved mysteries and unexplained phenomena, from the Kennedy assassination to the ominous black helicopters of the New World Order. In the sub-culture of the paranormal, his reputation was such that writers for "The X-Files" used to call him for ideas. In April 2001, fifteen hours after eating a meal with warm beef from a Mexican restaurant in Atlanta, he had agonizing evening of vomiting and diarrhea.

18   **Steven Mostow:** World's leading microbiologist dies of a sudden and suspicious death. On March 25, 2002, killed in a plane he was flying near Denver. He worked for the Colorado Health Sciences Centre,

19   **Guyang Huang:** World's leading microbiologist dies of a sudden and suspicious death. On Feb. 28, 2002, shot Tanya Holzmayer as she took delivery of a pizza and then apparently shot himself.

20   **Tanya Holzmayer:** World's leading microbiologist dies of a sudden and suspicious death. On Feb. 28, 2002, San Francisco resident was killed by a microbiologist colleague.

21   **Ian Langford:** World's leading microbiologist dies of a sudden and suspicious death. On Feb. 14, 2002, was found partially naked and wedged under a chair in Norwich, England.

22   **Victor Korshunov:** World's leading microbiologist dies of a sudden and suspicious death. On Feb. 9, 2002, had his head bashed in near his home in Moscow.

23   **Nguyen Van Set:** World's leading microbiologist dies of a sudden and suspicious death. On Dec. 14, 2001, died in an airlock filled with nitrogen in his lab in Geelong, Australia.

24   **Robert M. Schwartz:** World's leading microbiologist dies of a sudden and suspicious death. On Dec. 10, 2001, was stabbed to death in Leesberg, Va. Three Satanists have been arrested.

25   **Stanley Kubrick:** Movie producer of "Eyes Wide Shut" was killed by advanced technology. He also produced an earlier movie "2001 a Space Odyssey".

26   **Vladimir Pasechnik:** World's leading microbiologist dies of a sudden and suspicious death. On Nov. 21, 2001, a former high-level Russian microbiologist who defected in 1989 to the U.K. apparently died from a stroke.

27   **Don C. Wiley:** World's leading microbiologist dies of a sudden and suspicious death. Nov 16, 2001 he was missing and was found Dec. 20. Investigators said he got dizzy on a Memphis bridge and fell to his death in a river.

28   **Benito Que:** World's leading microbiologist dies of a sudden and suspicious death. On Nov 12, 2001, it was said he was beaten in a Miami parking lot and died later.

29   **Jim Keith:** Died in 1999. Author of many books including Mind Control, World Control. Jim died in hospital during surgery to repair a broken leg. A blood clot during the surgery traveled to his heart causing a pulmonary edema.

30   **Dean Stonier:** Organizer and promoter of the Global Sciences Congress. He hosted many top researchers including Phil Schneider and Al Bielek, the sole survivor of the Philadelphia Experiment. Dean died of a heart attack in August 2001, just a few months after a Denver Global Sciences Congress.

31   **Milton William Cooper:** Radio talk show host and author of the classic book: "Behold a Pale Horse" was shot dead during a gun battle with Sheriff deputies outside his home in Eager, Arizona. He believed that the government was not only concealing the existence of UFOs, but that they were working with the Illuminati to create a New World Order.

32   **Coral Lorenzen:** UFO researcher purportedly died of cancer.

33   **Ivan T. Sanderson:** A biologist died from rare brain cancer. He was head of a major UFO- paranormal group.

34   **Dr. Olavo Fontes:** Brazilian researcher. UFO researcher purportedly died of cancer.

35   **Gloria Lee (Byrd):** An author, in 1965 she said space people had told her to go on a fast during which she purportedly died of suicide.

36 **Jim Lorenzen:** UFO researcher purportedly died of cancer.

37 **Rev. Della Larson:** A contactee, she claimed that Venusians were living on earth among us. She purportedly died of suicide in 1965 in a rest home by hanging herself with a nylon stocking.

38 **Marie Ford:** A UFO enthusiast who discovered Larson's body, also had a suicide death

39 **Doug Hancock:** A UFO researcher purportedly died of suicide. He managed to obtain a gun and shot himself.

40 **Feron Hicks:** A UFO researcher purportedly died of suicide.

41 **Wilbert B. Smith:** Canadian researcher, UFO researcher purportedly died of cancer.

42 **Dr. Olavo Fontes:** Brazilian researcher. UFO researcher purportedly died of cancer.

43 **Jim Lorenzen:** UFO researcher purportedly died of cancer.

44 **Dr. B. Noel Opan:** Had cited UFO in 1959 and disappeared after an alleged visit by MIBs (Men in black) and was never seen again.

45 **Edgar Jarrold:** The Australian UFOlogist, also disappeared in 1960.

46 **H. T. Wilkins:** An author, died of a heart attack. He was well-known for his two books on UFOs, Flying Saucers on the Attack and Flying Saucers Uncensored.

47 **Henry F Koch:** Publicity director of the Universal Research Society of America. He cited a UFO on April 3, 1966, and died mysteriously a few weeks later.

48 **Damon Runyon Jr.:** UFO writer committed suicide by jumping or pushed off a Washington D.C. bridge in 1968.

49 **George Adamski:** A contactee, died of a heart attack in 1965. He claimed to have seen a flying saucer land in southern California and said he had spoken to its pilot, a Venusian, in front of witnesses, including George Hunt Williamson (Williamson wrote several books about UFOs and disappeared mysteriously in 1965 while on an anthropological expedition to Peru).

50   **Frank Edwards:** The noted news commentator and prominent UFOlogist, died of an alleged heart attack on June 23, 1967. He received two letters and a telephone call that he would not live for the World UFO Conference.

51   **Arthur Bryant:** A prominent UFOlogist and contactee died the next day after Frank Edwards on June 24, 1967.

52   **Richard Church:** Chairman of CIGIUFO also died the next day after Frank Edwards on June 24, 1967.

53   **Willie Ley:** UFO ET space writer, well-known writer on rockets and astronautics, Ley wasn't directly involved in UFO-logy but he wrote about space travel. Flying saucers are space travelers. He also died the next day after Frank Edwards on June 24, 1967.

54   **Rep. Rouse:** Had been supporting Edwards in his campaign for Congressional attention to the UFO issue, died of a similar heart attack.

55   **Frank Scully:** An author, died of a heart attack the next day after Frank Edwards on June 24, 1967. Scully wrote the first significant book about UFOs - Behind the Flying Saucers - in which he mentioned the "little men" or alien humanoids, electro-magnetic powerplants of saucer, EM effects, and the Air Force's campaign to hide the truth about UFOs from the public, all "ridiculous" ideas that were later accepted.

56   **James Forrestal:** Former Secretary of Defense committed suicide by purportedly jumping out a hotel window.

57   **M.K. Jessup:** Deeply immersed in the relatively new UFO-phenomena problem died of a suicide.

58   **Dr. James McDonald:** Senior physicist, Institute of Atmospheric Physics and also professor in the Department of Meteorology at the University of Arizona. In the 60s he tried to convince Congress to hold serious, substantial subcommittee meetings to explore the UFO reality. He died in 1971. He purportedly shot himself in the head. But it only was a gunshot wound to the head. He was wheelchair-ridden but somehow, several months later he drove to the desert and died of a fatal shot.

59 Dorothy Kilgallen: Dorothy was deeply immersed in the relatively new UFO-phenomena problem died in an alleged accident. A "special hush-hush meeting of the world's military heads" scheduled to take place the following summer. Received a 1955 dispatch, which barely preceded her death from an alleged overdose of sleeping pills and alcohol. She quoted an unnamed British official of cabinet rank, "We believe, on the basis of our inquiry thus far, that saucers were staffed by small men-probably under four feet tall. It's frightening, but there is no denying the flying saucers come from another planet."

60 Dr. Morris K. Jessup: An astronomer and archaeologist who was the first to reveal details of the top-secret Philadelphia Experiment - he died a few months later in 1959. He wrote, The Case for the UFO and The Expanding Case for the UFO, which had been depressed. The hose from his car exhaust was wired on; and it was, strangely, a washing machine hose.

61 add 30 Marconi deaths: Over 30-some-odd bizarre deaths associated with SDI (Star Wars) research at Marconi Ltd. in England between approximately 1985-1988. Below are six such deaths.

91 Roger Hill: A designer at Marconi Defense Systems, allegedly commits suicide with a shotgun, March 1985.

92 Jonathan Walsh: A digital communications expert employed by GEC, Marconi's parent firm, falls from his hotel room, November 1985, after expressing fear for his life.

93 Ashad Sharif: Another Marconi scientist, reportedly tied a rope around his neck, then to a tree, and in October 1986, got behind the wheel of his car and stepped on the gas with predictable results.

94 Capt. Don Elkin: Eastern Airlines pilot committed suicide. He had been investigating the UFO coverup for over 10 years and, at the time, was deep into the study of the Ra material with ('aria) Rucker. Reports of negative psychological interferences had developed during this latter investigation.

95   **Peter Ferry:** Marconi marketing director of the GEC firm, was found shocked to death with electrical leads in his mouth (August 1988).

96   **Alistair Beckham:** An engineer with the allied firm of Plessey Defense Systems was found shocked to death with electric leads attached to his body and his mouth stuffed with a handkerchief.

97   **Andrew Hall:** Another Marconi employee was found dead in September of 1988 of carbon monoxide poisoning.

98   **Coral and Jim Lorenzen:** The directors of APRO. A Tucson-based UFO group, both died of Cancer.

99   **Deck Slayton:** An astronaut was purportedly ready to talk about his UFO experiences, but cancer also intervened.

100   **Brian Lynch:** A young psychic and contactee, died in 1985, purportedly of a drug overdose. Did experimenting on psychic warfare techniques.

101   **Dr. James A. Hynek:** CUFOS founder, his death was because of "strange circumstances," died of rare brain tumor or cancer.

102   **Phil Schneider:** UFO investigator, died January 17, 1996 reportedly strangled by a catheter found wrapped around his neck. Of the 129 deep-underground facilities, Schneider believed the U.S. government had constructed since World War II, he claimed to have worked on 13. Claims that the American government concluded a treaty with "grey" aliens in 1954. This mutual cooperation pack is called the Grenada Treaty.

103   **Ron Johnson:** UFO investigator died on June 9, 1994. At the time MUFON's Deputy Director of Investigations. Johnson was 43 years old and, it would seem, in excellent health. Most recent job was with the Institute of Advanced Studies, purportedly working on UFO propulsion systems.

104   **Ron Rummel:** Ex-air force intelligence agent and publisher of the Alien Digest, died on August 6, 1993. Rummel allegedly shot himself in the mouth with a pistol. Friends say, however, that no blood was found on the pistol barrel and the handle of the weapon was free of fingerprints.

105   **Con Routine:** UFO investigator, like Phil Schneider and Ron Johnson also suspiciously died.

106   **Ann Livingston:** UFO investigator died in early 1994 of a fast-form of ovarian cancer. Wrote article, Electronic Harassment and Alien Abductions.

107   **Karla Turner:** UFO investigator died 1996. Wrote about UFO abductions.

109   **Danny Casolaro**: An investigative reporter looking into the theft of Project Promise software, a program capable of tracking down anyone anywhere in the world, died in 1991.

110   **Mae Brussell**: Acutely interested in UFOlogy. Many years of her research is available at The World Watchers Archive. A gutsy, no-holds-barred, investigative radio host died of a fast-acting cancer just like Ann Livingston and Karla Turner.

111   **Truman Bethurum:** He claimed to have ridden to other worlds in flying saucers. He was a quiet man who seemed incapable of making up the fantastic adventures he had. He wrote two books about them before dying quietly in bed on May 21, 1969.

112   **Dr. George Hunt Williamson:** Williamson disappeared while on an anthropological expedition to Peru in 1965. He was noted for his explorations of ancient Indian sites in the Andes, which he suspected were saucer bases, landing fields, and cave headquarters. He believed the saucermen were still there. Later, he experimented with shortwave radio contact, claiming in 1952, that he had established communications with UFOs. Saucers were observed hovering over his radio shack during these broadcasts. Williamson wrote several books about UFOs, the most noteworthy was The Road in the Sky.

BIBLIOGRAPHY

Armstrong, Karen (1993). A History of God, New York, N.Y., Ballantine Books.

Breasted, James H (1933), *The Dawn of Conscience*, New York, N.Y., Scribner's Sons.

Breasted, James H (1935), *A History of Egypt*, New York, N.Y., Scribner's Sons.

Bruno, Giordano (1950), *Giordano Bruno, His life and Thought*, New York, N.Y., Henry Schuman.

Budge, E.A. Wallis (1923), *The Book of the Dead*, London, England; Penguin Group.

Clayton, Peter A (1994), *Chronicle of the Pharaohs*, New York, N.Y., Thames and Hudson Inc.

Dimont, Max I (1962), *Jews, God and History*, New York, N.Y., Signet Books.

Freud, Sigmund (1967), *Moses and Monotheism*, New York, N.Y., Vintage Books div. of Random House.

Ginex, Nicholas P (2009), *Future of God Amen*, Bloomington, Indiana, Xlibris Corp. http://iranpoliticsclub.net/philosophy/amen/index.htm

Ginex, Nicholas P (2012), *AMEN, The Beginning of the Creation of God*, Bloomington, Indiana, Xlibris Corp.

Ginex, Nicholas P (2013), *Provide History of Religion and God*, Chute Institute, Littleton, Colorado, U.S. http://files.eric.ed.gov/fulltext/EJ1073192.pdf

Ginex, Nicholas P. (2013), *Allah, We, Our and Us*, Hancock Press, Fort Smith, AR.

Ginex, Nicholas P. (2018), *Understanding the Alien Extraterrestrial Mind.* http://iranpoliticsclub.net/features/extraterrestrial-mind/index.htm

Greer, Steven M. MD (2001), *2001 National Press Club Event.* https://www.youtube.com/watch?v=4DrcG7VGgQU

Greer, Steven M. MD (2009), Letter to President Obama, *The Center for the Study of Extraterrestrial Intelligence, The Disclosure Project* https://www.disclosureproject.org/docs/obama/obama-briefing-intro-letter.pdf

Holy Bible (1972), King James Version, *Gospel of John,* Red Letter Edition, Regency Publishing House, Nashville, Tennessee

Khan, Muhammad Zafrulla (1997), *The Qur'an*, Brooklyn, N.Y., Olive Branch Press (First published in 1970)

La Barre, Weston (1972), *The Ghost Dance*, New York, N.Y., Dell Publishing Co., Inc.

Manji, Irshad (2004), *The Trouble With Islam*, N.Y., N.Y., Canada, Random House, St. Martin's Press.

McCandlish, Mark (July 13, 2017) *Mark McCandlish: Master of Aerospace Illustration,* https://ufospotlight.wordpress.com/2017/07/13/mark-mccandlish-master-of-aerospace-illustration/

Noss, John B. (1974), *Man's Religions, 5th ed.,* New York, N.Y., MacMillan Publishing Co.

Pritchard, James B (1950, 1955), *Ancient Near Eastern Texts Relating to the Old Testament*, Princeton New Jersey, Princeton University Press, 2nd edition revised and enlarged

Pritchard, James B (1954, 1969), *The Ancient Near East in Pictures Relating to the Old Testament*, Princeton New Jersey, Princeton University Press, 2nd edition with supplement

Sagan, Carl (1980), *Cosmos*, New York, N.Y., Random House Inc.

Shaikh, Anwar (2016), Eternity Book Series, 2nd Ed, Book 4, Chapter 12, *Mind and Matter,* http://iranpoliticsclub.net/library/english-library/eternity4/index.htm

Smith, Homer W (1957), *Man and His Gods*, New York, N.Y., Grosset and Dunlap.

Spong, John S (2005), *The Sins of Scripture*, New York, N.Y., HarperCollins Publishers.

Steckling, Fred and Glen (2005), *We Discovered Alien Bases on the Moon,* translated from German to English in 2012, World Scientific Publishing Company

Tong, David (2017), *Quantum Fields: The Real Building Blocks of the Universe,* https://www.youtube.com/watch?v=zNVQfWC_evg

Tewari, Paramahamsa (2018), *Structural Relation Between the Vacuum Space and the Electron*, Physics Essays 31 (2018)

Tewari, Paramahamsa (2007), *Universal Principles of Space and Matter, A Call for Conceptual Reorientation* #10212 Astronomy (Books)

Vermes, Geza (1982), *The Dead Sea Scrolls in English*, Middlesex, England & N.Y., Penguin Books, Ltd.

Watson, John (2015), *Top Ten Scientific Flaws in the Big Bang Theory*, https://thetechreader.com/top-ten/top-ten-scientific-flaws-in-the-big-bang-theory/

DOES CONSCIOUSNESS PERVADE THE UNIVERSE?

Nicholas P. Ginex – April 5, 2018

Brief Summary

How did the universe begin? The most simplistic unit of all matter, the atom, is considered the basic building block of all matter. The hypothetical question arises, does the atom have as its goal, the development of inorganic and organic matter that can reach its height of perfection – to create the ability for thinking organisms to articulate a consciousness that can reach out and comprehend its own consciousness?

Start of Article

A novel idea has emerged which I would like our readers to share their thoughts. Consider the hypothetical idea that the unique positive and negative energy forces of an atom, which can coalesce into inorganic and organic matter, may have an inherent consciousness that tries to reveal itself. Ultimately, we, as thinking human beings, may be the product of that source of consciousness. We are part of the "stuff" that makes up the universe, and we are trying to understand our beginnings that could be due to the inherent forces of the atom that surfaces as consciousness. This hypothetical idea that matter, created by atoms that determine inorganic and organic outcomes, can assume consciousness may have some merit because we are proof as thinking products of our universe.

Another hypothetical idea that has surfaced is that our first awareness (consciousness) starts before conception through a transformation of energy into conscious awareness at birth. However, a question surfaces as to why only at birth of an organic form would consciousness come into play? I believe that the internal forces of the atom with its own negative, positive and neutral properties, has some level of consciousness that creates matter based upon its surrounding environment. Whether or not that matter is inorganic or organic will depend on the surrounding elements of heat, moisture, and an atmosphere conducive to life. Hence, for our earthly life forms (organic matter) to exist, the aggregate mix of atoms preordain with conditions that support life, the formation of different kinds of life forms.

Stated with earthly terms, plant and vegetable organic life begins first based upon inorganic matter (earth and minerals), heat, and moisture and it produces a byproduct (air) that sustains life. Organic forms of life that become mobile, such as the fly, a bird, fish, and animals are a consequence of the environment and it all began with the mix of atoms that produced inorganic and organic matter. So, I hypothesize that the atom has forces that somehow takes advantage of its surroundings and gives birth to inorganic and organic matter.

To me, it appears that such forces within the atom have an intelligence or a consciousness that tries to express itself in many ways. It is this consciousness that reaches its height in human beings to think and reach out with hypothetical thoughts to try to understand its own existence. And that is why I believe that the transformation of energy is a long, involved process whereby it transforms itself into matter that tries to express its inherent will to exist as consciousness.

Perhaps, that may be why gifted people have developed the idea of a soul because it is an extension of one's being with the hope that its soul will continue to exist even after death. Is this due to the internal energy of the atom that initiates the will for survival? It appears that forces within the atom, from its very beginning, fights to express itself in matter. So, I offer you the idea that the energy or forces inherent in the atom has a consciousness that seeks to find expression in the life forms it finally creates.

The question surfaces, what is this force or energy? Where did its positive, negative, and neutral forces come from? What gave this energy the will to exist as conscious entities that are very likely to appear in other parts of our universe? Are all life forms in other parts of the universe made of the same "stuff" that we humans are made of? I tend to believe "yes." The life forms on other planets may assume different shapes, colors, and organic functions that enable it to exist but they are all part of the energy that created the fundamental particles (electrons, protons and neutrons), which are part of the atoms that make up all the elements in the universe.

One last matter that needs to be addressed and that is can the internal forces of an atom be attributable to "God." I do not know because I have no valid conception of God. To me, God is mysterious, incomprehensible, and unknowable. However, I use the term God because it introduces a concept of the consciousness that many people feel within themselves, which is the ability to love. The energy or force of the atom has given life forms the will to survive and therefore the desire to love and reproduce itself and that is why the atom is part of every element, where from inorganic matter evolves organic, living life forms.

The next question I offer is, if you give this unknown force a name, such as God, why not put it to good use? That is, why not sublimate our conception of the unknown force into idealistic forms of moral behavior that allows people from different countries and of different races, to love and accept one another as sisters and brothers? This use of the unknown force is not a bad idea. The alternative is to say we are exceptionally bright and since we understand the unknown force, just except it, and let's live without any need for a moral code of behavior because we can control our lives without anybody telling us what we should or should not do. This attitude I cannot agree with because it negates the wisdom we have gained in the past through experience of what makes a harmonious society.

My concept of God does not assume any preexisting definitions of God and therefore does not present any myths or images because as I had stated before, as with the internal forces of the atom, God is mysterious, incomprehensible, and unknowable. The idea of God becomes a sublimation of our finest thoughts as to what mankind can aspire to in terms of integrity and love for one another.

Men have always wondered about how did life begin and if there is a God that created all there is? The Egyptians developed the greatest civilization with a belief in a creator god. After thousands of years, the Priesthood of Amon wrote the first documented scripture (a belief) titled, *Amon As the Sole God* during the reign of Ramses II. Readers who thirst for knowledge of how the concept of a soul developed, a Hereafter, a Son of God, and belief in one-universal God, they may access the book titled *Future of God Amen.* It is hosted on the link below as a free read.

http://iranpoliticsclub.net/philosophy/amen/index.htm

Do you agree with the hypothetical idea that the atom, after many transformations from inorganic to organic matter, strives to achieve a consciousness to express itself by ultimately producing life forms with the ability to think? Can it be possible that this consciousness pervades the universe and somehow embodies the essence of God?

EXPOSE THE QURAN WITH WORLDWIDE COMMUNICATION

Nicholas Ginex – February 12, 2018
This article also is presented on the following websites:
http://www.nicholasginex.com/2018/02/07/
stop-islamic-terrorism-by-exposeing-the-quran/
http://iranpoliticsclub.net/islam/expose-quran/index.htm

OVERVIEW

A worldwide problem is Islamic expansion. Muslims are taught that their religion is the religion of truth and shall prevail over all others (Qur'an 9:33). News media in Egypt, the U.S. and Europe, will someday find out that they too will be the targets of oppression as Islam prohibits freedoms to explore many avenues of free-independent thought. A SOLUTION is to EXPOSE the ideology of Islam presented in the Qur'an.

People around the world have got to communicate that the Qur'an is the root, the source of discontent and terrorism throughout the world. Knowledge of Islam's despicable history and abominable verses in the Qur'an will lead to an inevitable solution – an Islamic Reformation.

Stop Islamic terrorism With World-Wide Communication.

People must EXPOSE the Quran by communicating worldwide the many abominable verses thatare indoctrinated into unsus-pecting minds.

Worldwide person to-person communication requires the active participation by business leaders, humanitarians, media news outlets, government (Senate and House of Representatives), agencies (such as the CIA and FBI), members of forums and political clubs. Their combined efforts will *enable a viable solution* that *EXPOSES abominable verses in the Qur'an*.

Members of the Middle East Forum (MEF), Iran Politics Club (IPC), American Islamic Forum for Democracy (AIFD), the Islamic Board, and the Stand for Peace Islamic Forum of Europe (IFE) *are requested to help stop Islamic terrorism by exposing the Quran*. They have the media and financial resources to actively support communication efforts *to help Muslims understand the need to revise the Quran and attain a peaceful solution.*

Reporting of Islamic atrocities by writers and journalists in many countries keeps the public aware of the Islamic threat but that **is NOT enough**. Their efforts reveal Islamic extremism and its expansion but *does NOT eliminate radical Islam. There is a SOLUTION.*

A Solution to Eliminate Islamic Extremism

Communication efforts to **EXPOSE the Quran** are **NOT an attempt to incite anger and hate against Muslims or denigrate the religion of Islam**. It will allow Muslims an opportunity to *understand the flaws of their religion*. Through communication, Muslims will be able to understand that *a solution* is possible. If they *truly want to live in peace with non-Muslims* around the world, they will pressure their religious leaders and scholars to **revise the Quran.**

Communication Objectives to EXPOSE the Quran.

- **Inform Muslims worldwide *why the Quran* is the source of discontent in many countries and that *it is the root cause of Islamic extremism and terrorism.***

- *Muslims need to learn of Islam's despicable history.*

- **Open** *a dialog with Muslims to acknowledge abominable verses in the Quran* and *advocate God's greatest command, love one another.*

Lists of *abominable verses in the Qur'an* are provided in Section 3.3 of *Allah, We, Our and Us* and by Zulfikar Khan in his article, *Islamic Hell Torture Chamber for Unbelievers*. The links are respectively:

http://iranpoliticsclub.net/library/english-library/allah-we/index.htm

http://iranpoliticsclub.net/islam/islamic-hell/index.htm

Only informed Muslims can initiate an Islamic Reformation.

WHY People Worldwide Must Communicate.

In 100% of mosques, imams, caliphs and mullahs teach their Muslim followers to believe the Quran is the literal word of God and Islam will prevail over all other religions (Quran 9:33). Most Muslims are illiterate and indoctrinated with Quran beliefs from childhood and earlier. They are unaware that the Quran contains many abominable verses that incites bigotry, hatred, violence, and the murder of non-Muslims. To fully understand *how* the Muslim mind is indoctrinated, the following link is provided. It presents, *Indoctrination of the Muslim Mind*.

http://iranpoliticsclub.net/islam/indoctrination/index.htm

It reveals *why* it is essential to learn of the insidious psychological use of "We, Our and Us" in the Quran. The plural pronouns for Allah *creates for Muslims* –

a feeling of unity with the Creator and the compulsion to impose this unity on all people with the altruistic belief that they are promoting the integration of mankind with the Creator.

History Reveals Islam Was Never a Religion of Peace

People need to communicate worldwide the *despicable history of Islam*. This religion grew by having Islamic armies capture the lands of many countries and their captives were given two alternatives: convert to Islam and pay a tax or die by the sword. It continues to expand in many countries as Islamic followers are not able to assimilate with people with other beliefs and many, in response, commit acts of terrorism. A complete history of the rise of Islam and an objective presentation of the many abominable verses in the Qur'an is presented in, *Allah, We, Our and Us*. It is hosted on:

http://iranpoliticsclub.net/library/english-library/allah-we/index.htm

This book was written to inform people *why* Islam
is a danger to all people around the world and to
communicate *why* the *Qur'an must be revised*.

*If Muslims Fail to Expose and Delete
Abominable Verses in the Qur'an,
Islamic Extremism and Terror Will
Always Exist.*

In a Gatestone Institute article dated February 7, 2018 titled, *Death of Democracy? - Part I*, Denis MacEoin wrote,

"Rather than stand up to our enemies, both external and internal, are we now so afraid of being called "Islamophobes" that we will sacrifice even our own cultural, political, and religious strengths and aspirations?"

Unafraid, we can communicate openly and respond to the stoning of an Iranian woman on the cover of *Allah, We, Our and Us – Her agony and pain provokes compassion, justice and truth.*

Silence Permits an Islamic Cancer to Grow

The solution provided by worldwide communication presents an opportunity for **all people** to help solve the looming threat of Islam *by identifying the problem* and **understanding** *why the Qur'an must be revised. Communication is an alternative to avert* the eventuality of *WWIII.*

A silent response signifies fear, unable to confront the enemy that kills millions of people in the name of an Islamic God. It is time to *'Wake Up' or continue to permit Islam to expand and eventually dominate the values, culture, and laws of every country.*

It is appropriate to end this appeal for all people to communicate in *EXPOSING the Quran* with a song sung by Paul Simon and Art Garfunkel. The song poignantly **reveals the** *TRUTH* **about Islamic terrorism.** The video illustrates the hate, violence and death caused by Muslim extremists that revere the sanctity of the Qur'an.

https://www.youtube.com/watch?v
https://www.youtube.com/watch?v=BsKGaWS3sGE
https://vimeo.com/161554060

NOTE: The above YouTube and Vimeo videos have been taken off the Internet for politically correct reasons. It is a realistic example of Media Control in the U.S. by the **Surreptitious Shadow Government (SSG)** identified in **Everything Has a Beginning – Even the Universe,** under *3.0 Disclose Zero-Point Energy for Mankind.*

The appeal is directed to Imams, Caliphs, Mullahs and their followers to call for an Islamic Reformation to revise the Qur'an with verses that will guide Muslims followers to love all people and respect all faiths. Failure to do so will cause terrorism to always exist.

Education is Key for an Islamic Reformation

Education is key for Muslims to succeed in revising the Quran. This is possible through **Worldwide Communication** between Muslims and people in countries around the world.

Educated, intelligent people of Iran cannot continue to tolerate the repression of their individual freedoms by an archaic ideology founded on the Quran and enforced by Sharia law.

Muslims must, if not educated, be informed to understand that they must change a flawed ideology that kills innocent people in the name of God. By people around the world flooding the Internet with information that exposes the hatred, bigotry, and violence contained in the Qur'an, the flawed ideology of Islam can be revised.

Education through Communication will benefit Muslims to understand the threat of Islam and energize them *to revise the Quran*. People who are willing **to *EXPOSE* the Quran**, will be helping *Muslims to help themselves.*

Worldwide communication averts retaliation, such as dropping bombs on Muslims in occupied territories to eliminate Islamic repression. Rather, it can be achieved by dropping copies of *Allah, We, Our and Us* into Muslim and non-Muslim territories so *that they become knowledgeable of the Islamic threat and a viable solution.*

Royalties for *Allah, We, Our and Us* will be used to **Expose the Quran** in the **Worldwide Communication effort**. It is provided as a Free Read via:

http://iranpoliticsclub.net/library/english-library/allah-we/index.htm

An Islamic Revolution by an Egyptian President

Worldwide Communication will enable Muslims to effectively mobilize and achieve a successful reformation movement expressed by Egypt's President. On January 1, 2015, President Abdel Fattah el-Sisi spoke to Islamic scholars from Al-Azhar, the highest center of Sunni Muslim learning. His message was for Muslim leaders to rethink Islamic discourse and prevent it from *"antagonizing the entire world!"* This outstanding president is not only a pious Muslim but a perceptive intellectual leader who perceives that, *"We need a revolution of the self, a revolution of consciousness and ethics to rebuild the Egyptian person."*

Parts of his impassioned speech are provided below:

"I say and repeat, again, that we are in need of a religious revolution. You imams are responsible before Allah. The entire world is waiting on you. The entire world is waiting for your word ... because the Islamic world is being torn, it is being destroyed, it is being lost. And it is being lost by our own hands,"

"We need a revolution of the self, a revolution of consciousness and ethics to rebuild the Egyptian person — a person that our country will need in the near future,"

"It's inconceivable that the thinking that we hold most sacred should cause the entire Islamic world to be a source of anxiety, danger, killing and destruction for the rest of the world. Impossible that this thinking — and I am not saying the religion — I am saying this thinking," He continued: *"This is antagonizing the entire world. It's antagonizing the entire world! Does this mean that 1.6 billion people (Muslims) should want to kill the rest of the world's inhabitants — that is 7 billion — so that they themselves may live? Impossible!"*

President el-Sisi alluded to, but did not specifically identify the *"source of anxiety, danger, killing and destruction for the rest of the world."* Indirectly, he said, *"this thinking"* that Muslims hold as most sacred *"is antagonizing the entire world."* That **sacred source**, indoctrinated into the Muslim thinking mind, is the **Qur'an**. It is quite plausible that religious leaders and scholars listening to President el-Sisi knew he was referring to the Qur'an as the sacred source.

An Islamic Revolution can initiated by enlightened Muslims that understand the need to revise the Quran and advocate God's greatest command – Love one another.

AUTHOR BIO

Nick enjoys his 71ˢᵗ birthday August 27, 2006
Photographer: Jennifer Schwartz

Nick Ginex received a B.E.E degree at the City University of New York and an MBA in Finance at Adelphi University in NY. He worked in several aerospace companies in the support disciplines of Maintainability, Human Factors and Configuration Management (CM).

As CM Manager of software and hardware products for top aerospace and commercial companies, his planning and organizational skills were applied for the successful operation of entire engineering projects.

While writing his first book about the history of religion and God he sang and played his guitar at senior care centers and nursing homes for their enjoyment. The smiles on their faces and the joy in their eyes have been his greatest reward.

Nicholas Ginex books have been written to inform and educate people to understand how mankind came to conceive a God. A God that has influenced the development of the Judaic, Christian and Islamic religions. He was motivated to write a paper titled, ***Provide***

History of Religion and God. It was featured on the Internet by ERIC (Education, Research and Information Center) and resides on:

http://files.eric.ed.gov/fulltext/EJ1073192.pdf

Nicholas gravitated from theological interests into the scientific-philosophical realm and the political arena. As an author, he has been interviewed on a number of radio shows including the renowned international 'X' Zone Radio Show hosted by Rob McConnell. His peers and associates regard him as a free and creative thinker with perceptive ideas that allow them to question and revisit many aspects of their theology and philosophical views.

There are overviews of eight novels that reside on the website,

http://www.futureofgodamen.com

This site provides an e-mail capability. However, readers may correspond from any computer by using Nicholas' direct e-mail address:

nickginex@gmail.com

Ginex articles are hosted on,

http://www.nicholasginex.com

The following commendation was provided by Dr. Ahreeman, the creator and administrator of IPC, the Iran Politics Club.

Nicholas Ginex is a valuable philosopher, Egyptologist, scholar, author, and a scientific person. Inside IPC (Iran Politics Club), he needs no introduction and outside IPC, all you need to know is that he is a true American Patriot, Concerned Global Citizen and a Lovely Humanitarian. Nicholas is well educated and radically logical. He seeks the truth, exposes corruption and offers solutions. His writings are strongly recommended to read.

Dr. Ahreeman X, IPC Founder

Dr. Ahreeman, an American of Persian background, created his IPC website to preserve many of the greatest Persian writers, inventors, painters, poets, and philosophers. There, he has honored Mr. Ginex by posting 3 of his books, many articles and his most recent book, *Artificial Intelligence Can benefit Mankind*. All four books may be accessed by going to

http://iranpoliticsclub.net/authors/nicholas-ginex/index.htm

This book, *Everything Has a Beginning – Even the Universe* may also be hosted on Dr. Ahreeman's website. All five books may be accessed by going to the Nicholas Ginex index link above.

A scientific and logical thinker, in addition to writing articles about the beginning of the universe, the phenomenon of consciousness, the reality of extraterrestrials, and the need to prepare mankind for many marvelous journeys into the universe, Nicholas seeks to find balance in the economy of the United States. He has presented in the release of a timely book, *Artificial Intelligence Can benefit Mankind*, a simple, yet innovative solution that can restore sanity in our conception of net worth for all people that work and sustain the greatest country on earth – the United States.

www.ingramcontent.com/pod-product-compliance
Lightning Source LLC
Chambersburg PA
CBHW021448210526
45463CB00002B/685